FLORA ZAMBESIACA

Flora terrarum Zambesii aquis conjunctarum

T0132981

VOLUME SEVEN: PART FOUR

FLORA ZAMBESIACA

MOZAMBIQUE

MALAWI, ZAMBIA, ZIMBABWE

BOTSWANA

VOLUME SEVEN: PART FOUR

Edited by
E. LAUNERT & G. V. POPE

on behalf of the Editorial Board:

G. Ll. LUCAS
Royal Botanic Gardens, Kew

E. LAUNERT
The Natural History Museum

M. L. GONÇALVES
*Centro de Botânica, Instituto de Investigação
Científica Tropical, Lisboa*

G. V. POPE
Royal Botanic Gardens, Kew

Published by the Managing Committee on behalf of
the contributors to Flora Zambesiaca
1990

Typeset at the Royal Botanic Gardens, Kew, by
Pam Arnold, Christine Beard, Brenda Carey,
Margaret Newman, Helen O'Brien
and Pam Rosen

Printed in Great Britain by
Whitstable Litho Printers Ltd.,
Whitstable, Kent

ISBN 0 950 7682 7 8

CONTENTS

LIST OF FAMILIES INCLUDED IN
VOLUME VII, PART 4

6A. CANELLACEAE

By B. Verdcourt

Aromatic glabrous trees. Leaves simple, alternate, gland-dotted, without stipules. Inflorescences axillary or axillary and terminal, cymose. Flowers regular, hermaphrodite. Sepals 3, persistent. Petals 5–12, free or united into a tube at the base, rather thick, in 1 or 2 whorls. Stamens hypogynous, 6–12; filaments united into a tube; anthers extrorse, adnate to the upper part of the tube, bilobed, each lobe composed of two microsporangia. Ovary superior, unilocular; ovules 2–many on 2–6 parietal placentas, subanatropous. Style short and thick; stigmas 2–6. Fruit a berry. Seeds 2–many, shining with oily fleshy endosperm.

A family of 5 genera and 17 species in tropical North and South America, east tropical Africa and Madagascar. Only *Warburgia* occurs in Africa. The sepals, inner and outer petals have also been considered to be bracts, sepals and petals respectively but Wilson (see below) and Parameswaran (Proc. Indian Acad. Sci., Sect. B, **55**: 167–182 (1962)) show that the vascularisation suggests the former interpretation.

WARBURGIA Engl.

Warburgia Engl., Pflanzenw. Ost-Afr. **C**: 276 (1895). —Burtt Davy, Fl. Transv. **1**: 218 (1926). —Verdc. in F.T.E.A., Canellaceae: 1 (1956). —Melchior & Schultze-Motel in Pflanzenfam. ed. 2, **17a**, II: 223 (1959). —Hutch., Gen. Fl. Pl. **1**: 64 (1964). —Verdc. in Regnum Vegetabile **40** (Nom. Conserv. Prop. II): 27 (1965). —Wilson in Am. Journ. Bot. **53**: 336 (1966); in Taxon **17**: 328 (1968). —Codd in Fl. S. Afr. **22**: 40 (1976) nom. conserv. *Chibaca* Bertol.f. in Mem. Accad. Sci. Istit. Bologna **4**: 545 (1853); Illustrazione de piante Mozambicesi, Dissertazione **3**: 13. t. 3 (1854).

Evergreen trees. Leaves very shiny above. Inflorescences axillary, the flowers solitary or 3–4 in reduced cymes. Sepals broadly ovate to reniform, very obtuse. Petals 10 in 2 rows, the outer larger and thicker than the inner, obovate-spathulate, obtuse, narrowed to the base, pellucid gland-dotted. Stamens 10, united into a tube, the part extending above the anthers 10-crenellated; anthers 10. Ovary narrowly ellipsoid, narrowed toward the apex, clavate at the base; ovules 10–31 in 1–2 rows on 5 placentae. Style obconic, truncate at the apex, bearing 5 oval stigmatic patches round the side and obscurely 5-lobed at the apex. Fruit a globose, ovoid or narrowly ellipsoidal berry with coriaceous pericarp; seeds immersed in green pulp.

A small genus of 4 species, 3 closely related, the other diverse; occurring in E. Zaire, S. Ethiopia, E. Africa, Malawi, Zimbabwe, Mozambique and NE. South Africa.

Fruit 2–3 × 1.5–2.5 cm. often rather small; ovules 15–20 - - - - - 1. *salutaris*
Fruit up to 5 cm. in diam.: ovules 25–30 - - - - - - - 2. *ugandensis*

1. **Warburgia salutaris** (Bertol.f.) Chiov. in Nuov. Giorn. Bot. Ital. n. ser. **44**: 683 (1937). —Mendes in Bol. Soc. Brot. Sér. 2, **43**: 338 (1969). —Ross, Fl. Natal: 247 (1972). —Palmer & Pitman, Trees S. Afr. **3**: 1523 (1972). —Codd in Fl. S.Afr. **22**: 40, fig. 10 (1976). —Palgrave, Trees S. Afr.: 618 (1977). Tab. **1**. Type: Mozambique, Delagoa Bay, *Fornasini* s.n. (BOLO, holotype†).
    Chibaca salutaris Bertol.f. in Mem. Accad. Sci. Istit. Bologna **5**: 545, t. 23 (1853); Illustrazione de piante Mozambicesi, Dissertazione **3**: 13, t. 3 (1854). Type as above.
    Warburgia breyeri Pott in Ann. Trans. Mus. **6**: 60, fig. 1, 2 (1918). —Burtt-Davy, Fl. Transv. **1**: 218, fig. 32 (1926). Type from S. Africa (Transvaal).

Tree or shrub 4–10(18) m. tall with rough mottled grey bark, red on inner side; branches striate and lenticellate. Leaves 4.5–11 × 1.5–2.5 cm., oblong to oblong-lanceolate or elliptic, acute at the apex, cuneate at the base, shiny above, paler beneath, densely pellucid-dotted; petiole 2–5 mm. long. Flowers solitary or in short 3-flowered cymes; peduncle 2–3 mm. long; bracts 0.5 mm. long, deciduous leaving conspicuous scars;

1

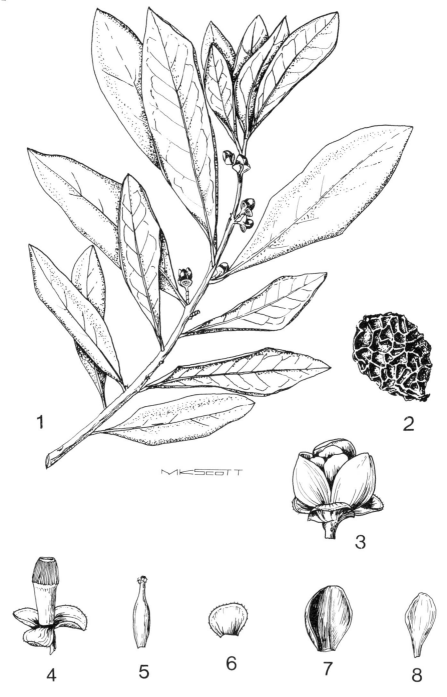

Tab. 1. WARBURGIA SALUTARIS. 1, flowering twig (× 1), from *Strey & Schlieben* 8611; 2, fruit (× 1), from *Van Warmelo* s.n.; 3, flower (× 4); 4, staminal column (× 7); 5, ovary (× 7); 6, sepal (× 4); 7, outer petal (× 4); 8, inner petal (× 4), 3–8 from *Strey & Schlieben* 8611. From Flora of Southern Africa.

pedicels c. 1.5 mm. long. Sepals 2 × 3 mm., minutely ciliate. Petals: outer 4–5 × 3 mm., obovate, concave, subcoriaceous; inner yellow, 4 × 1.5 mm. spathulate, thinner. Staminal tube 3–4 mm. long; anthers 1.5 mm. long. Ovary 3 mm. long; ovules 15–20. Fruits reddish-blue, 2–3 × 1.5–2.5 cm.; pericarp coriaceous, wrinkled; seeds several, flattened.

Zimbabwe. E: Chipinge Distr., Tanganda R., 8 km. N. of Tanganda Tea Estate, Mihizu, st. 6.viii.1973, *Mavi* 1467 (SRGH). **Mozambique**. M: Namaacha, Rd. towards Matianine, fl. 31.iii.1968, *de Carvalho* 1000 (LISC; LMA).

Also in S. Africa (Natal and Transvaal); cultivated in Harare Botanic Garden. Lowland evergreen forest under 200 m. in Maputo but also at c. 1000 m. in *Blighia-Melanodiscus* association in Chipinge District.

The bark has been used medicinally since early times against colds, chest complaints and malaria. The present rarity of the tree is probably due to the strong demand; small pieces fetch a relatively high price. In Maputo District known as Chibaha (Xibahá) hence Bertolini's original generic name.

2. **Warburgia ugandensis** Sprague in Journ. Linn. Soc., Bot. **37**: 498 (1906). —Dale & Eggeling, Indig. Trees Uganda ed. 2: 71, col. pl. 5 (1952). —Verdc. in Kew Bull. **9**: 543 (1955); in F.T.E.A., Canellaceae: 2, fig. 1 (1956). —Dale & Greenway, Kenya Trees & Shrubs: 111, col. pl. 10, phot. 25 (1961). —Boutique in F.C.B., Canellaceae: 2, fig. 1 (1967). Type from Uganda.

Tree 5–42 m. tall with rough fissured dark brown or brownish grey bark dividing into rectangular plates; slash often dark red. Leaves 3–15 × 1.4–5 cm. oblong-lanceolate, elliptic or oblong-elliptic, often a little falcate, acute at the apex, cuneate at the base, very shiny and dark green above, paler beneath, densely pellucid-dotted; petiole 3–5 mm. long. Flowers solitary or in short 3–4-flowered cymes and sometimes with a single flower from the base of the cyme, the true peduncle then being obsolete; peduncle 3 mm. long; bracts ± square, 2 mm. long, very deciduous leaving conspicuous scars; pedicels 3–4 mm. long. Sepals thick, 3 × 3–3.5 mm., ciliate. Petals: outer 6–7 × 4–4.5 mm.; inner 5–7 × 2.5–3 mm., obovate-spathulate. Staminal tube 4–5 mm. long; anthers 1.5–2 mm. long. Ovary 2.6–4 mm. long; ovules 25–30. Fruit up to 5 cm. in diam., at first greenish and ellipsoidal, later subspherical, turning purplish or with blue-grey bloom when ripe; seeds 1–1.5 cm. long, yellow-brown, compressed, ± cordate.

Subsp. **ugandensis**

Tree up to 42 m. tall. Leaves oblong-elliptic or elliptic, 3–15 × 1.4–5 cm. Sepals 3 × 3.5 mm. Petals, outer 6–7 × 4–4.5 mm., inner 5.5–7 × 3 mm. Staminal tube 5 × 2–3 mm.; anthers 2 mm. long. Ovary 4 mm. long; style 1 mm. long; ovules 25–30.

Malawi. N: Nkhata Bay Distr., S. Viphya Forest Reserve, near Zumwanda Compound, fr. 26.vii.1982, *Chapman* 6303 (K). S: Mt. Mulanje, Likhubula Valley at source of Nasato Stream below Cilemba Cliffs, fr. 2.x.1986, *J.D. & D.G. Chapman* 8111 (K; MO).

Also from Zaire, Ethiopia, Uganda, Kenya and Tanzania. Evergreen forest and *Tecomaria-Buddleja-Dodonaea* bracken thicket with scattered trees; 1200–1700 m.

I formerly included *Warburgia breyeri* (i.e. *salutaris*) in the synonymy of *Warburgia ugandensis* (F.T.E.A., Canellaceae: 3 (1956)) but I am prepared to follow Codd in keeping them separate; there are small but definite differences in fruit size and ovule number.

At first I considered *Chapman* 6303 to belong to subsp. *longifolia* (Verdc. in Kew Bull. **9**: 543 (1955); in F.T.E.A., Canellaceae: 3 (1956)) described from the Rondo Plateau in SE. Tanzania, but am now satisfied that the few trees in N. & S. Malawi are closer to the more northern populations in East Africa.

Subsp. *longifolia*, now known from five gatherings all from the Rondo Plateau, has narrower more parallel-sided leaves with more distinct petioles. Flowers are needed for further investigation of the Malawi plants.

113. GENTIANACEAE

By J. Paiva and I. Nogueira

Annual or perennial herbs, sometimes suffrutescent, rarely shrubs or small trees, erect to straggling, terrestrial or aquatic, rarely parasites or saprophytes, usually glabrous. Leaves simple, opposite (in the Flora Zambesiaca area), or seldom whorled, only rarely alternate, exstipulate, sometimes reduced and scale-like. Inflorescence a terminal cyme, often paniculate, or axillary and fasciculate or, sometimes, reduced to solitary flowers.

Flowers hermaphrodite or sometimes unisexual, actinomorphic or rarely zygomorphic, subsessile or pedicellate. Calyx tubular of (2)4–5, or rarely 6–12 sepals (lobes), imbricate (sometimes valvate). Corolla gamopetalous; tube campanulate, funnel-shaped or cylindric, sometimes with a constricted limb, 3–5(12)-lobed, usually contorted or rarely imbricate, often with scales or nectary-pits within the tube. Stamens as many as the corolla lobes, alternating with them, inserted on the corolla; filaments usually dilated at the base; anthers 2-thecous, with distinct parallel cells dehiscing lengthwise, sometimes twisted, or seldom with apical pores. Disk obsolete or annular or of 5 hypogynous glands. Ovary superior, uni- or bilocular (through the intrusion of the cell walls), usually with many ovules in each cell, with parietal or, in the bilocular ovaries, axile placentation; style simple or rarely absent; stigma entire or bilobed or divided into 2 filiform branches (rarely stigmas deccurent along the sides of the ovary when style wanting). Fruit usually capsular and dehiscent septicidally bivalved or rarely berry-like and splitting irregularly. Seeds usually numerous, subglobose, polyhedric to ovoid-ellipsoid, smooth, wrinkled, or reticulate-faveolate or sometimes frilled; testa crustaceous or membranous; embryo small, straight and elongate, embedded in the copious oily albumen.

A family of c. 75 genera and about 1000 species, cosmopolitan but most common in temperate and subtropical regions and in tropical mountains.

1. Ovary bilocular through the intrusion of the cell-walls, the placentas eventually appearing axile - - - - - - - - - - - - - - - - 2
– Ovary unilocular with parietal or sometimes intrusive placentas - - - - - 3
2. Flowers mainly yellow, sometimes cream or white; corolla tube cylindrical (up to 30 mm. long) or funnel-shaped (3–6 mm. long); anthers opening by longitudinal slits, without or with 1 apical gland, and without or with 1–2 basal glands, very rarely with only 2 basal glands - - - - - - - - - - - - - - - **4. Sebaea**
– Flowers mainly blue, rarely pink or white; corolla tube short (2.5–6.5 mm. long), campanulate or broadly cylindrical; anthers opening by terminal pores later prolonged opening into longitudinal slits during the dehiscence, without apical and basal glands - - - - - - - - - - - - - - - **3. Exacum**
3. Flowers 3-merous - - - - - - - - - - - **10. Pychnosphaera**
– Flowers 4–5-merous - - - - - - - - - - - - - - 4
4. Corolla lobes with 1–2 basal glandular nectaries within - - - - **5. Swertia**
– Corolla lobes without nectaries - - - - - - - - - - - 5
5. Flowers (4)5-merous - - - - - - - - - - - - - - 6
– Flowers typically 4-merous - - - - - - - - - - - - - 7
6. Flowers in axillary clusters; calyx with 3 short basal glands; filaments with a double-hooded scale at the base - - - - - - - - - - - - - **11. Enicostema**
– Flowers in lax cymes, sometimes solitary; calyx without basal glands; filaments without scales at the base - - - - - - - - - - - - - - **6. Chironia**
7. Flowers axillary, solitary, 2–3-nate, or in terminal racemes or corymb-like clusters - - - - - - - - - - - - - - - - 8
– Flowers in terminal or axillary few to many-flowered cymes - - - - - 9
8. Calyx leathery; lobes 3-nerved - - - - - - - - **1. Neurotheca**
– Calyx papyraceous; lobes 1-nerved - - - - - - - **2. Congolanthus**
9. Corolla tube with 4 more or less semilunate, fimbriate-papillose scales; stamens subequal, all fertile - - - - - - - - - - - - - - **9. Faroa**
– Corolla tube without scales; stamens unequal, 1–3 sterile, rarely all fertile - - - 10
10. Superior leaves much reduced and scale-like; flowers in compact terminal clusters; corolla lobes as long as the tube and more or less equal - - - - - - - **8. Schinziella**
– All leaves alike; flowers in lax terminal or axillary cymes, sometimes solitary; corolla lobes shorter than the tube, usually unequal, 2 larger and 2 shorter - - - **7. Canscora**

1. NEUROTHECA Salisb. ex Benth.

Neurotheca Salisb. ex Benth. in Benth. & Hook. f., Gen. Pl. **2**: 812 (1876). —Raynal in Adansonia, Sér. 2, **8**: 57 (1968).

Annual or perennial herbs. Stems erect, slender, simple or branched at the apex. Basal leaves elliptic; stem leaves from elliptic-ovate (lower) to linear or linear-lanceolate (upper). Flowers 4-merous, axillary, in long or short racemes or crowded in terminal corymb-like clusters, rarely solitary. Calyx leathery, tubular, 8–12 ribbed; lobes 3-nerved. Corolla tube cylindric or narrowly funnel-shaped, with 4-ovate lobes. Stamens 4;

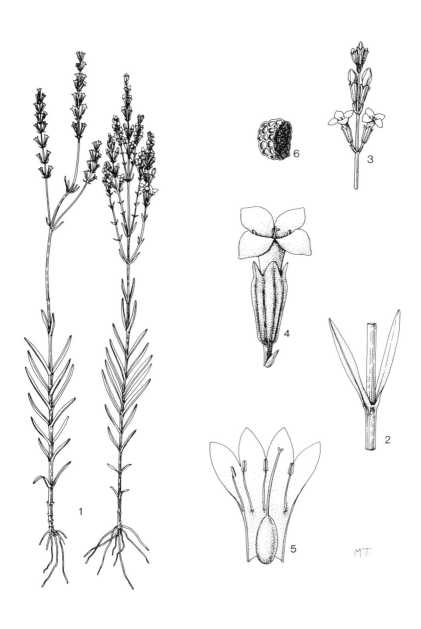

Tab. 2. NEUROTHECA CONGOLANA. 1, habit, plant on left fruiting, plant on right flowering (×½); 2, stem node, leaves opposite (× 1); 3, part of inflorescence (× 1); 4, flower (× 4); 5, corolla opened out to show pistil and stamens (× 4); 6, seed (× 20), all from *Mendonça* 2482.

filaments inserted below the middle of the corolla tube, filiform, slightly unequal; anthers oblong-ellipsoid, not twisted. Ovary oblong-ovoid, unilocular; ovules numerous; style filiform; stigma bilobed. Capsule membranous, oblong-obovoid, bivalved.

A small tropical African genus with 3 species.

Neurotheca congolana De Wild. & Th. Dur. in Bull. Soc. Roy. Bot. Belg. **38**: 98 (1899). —Baker & N.E. Br. in F.T.A. **4**, 1: 562 (1903). —Raynal in Adansonia, Sér. 2, **8**: 66, t. 4 fig. 1–2 (1968). —Boutique in Fl. Afr. Centr., Gentianaceae: 16 (1972). TAB. **2**. Type from Zaire.
    Neurotheca baumii Gilg in Warb., Kunene-Samb.-Exped. Baum: 333 (1903). Type from Angola.
    Neurotheca schlechteri Gilg ex Baker in F.T.A. **4**, 1: 561 (1903). —Ross in Bot. Surv. Mem. 39: 279 (1972). Type: Mozambique, Inhambane, *Schlechter* 12087 (COI; FI; K, holotype; P).
    Octopleura loesilioides var. *grandiflora* Knoblauch in Bot. Zentralbl. **60**: 362 (1894) pro parte quoad specim. *Buttner* 475. Syntypes from Sierra Leone and Zaire.

Annual herb 20–30 cm. tall, usually unbranched. Stem slender glabrous. Basal leaves sometimes crowded, 7–10 × 2–4 mm., elliptic-ovate, shortly petiolate; stem leaves opposite, 3–20 × 1–3.5 mm., linear or linear-lanceolate, sessile. Flowers blue or whitish, in terminal racemes; pedicel 1.5–3 mm. long, enlarged in fruit; bracts 4–5 mm. long, linear. Calyx tube 4–5 mm. long, equally and prominently 8-ribbed; lobes 1.5–2 mm. long, deltoid, acute. Corolla tube, 6–7 mm. long; lobes 1.5–3 mm. long, broadly ovate, obtuse. Filaments c. 3 mm. long, linear; anthers 0.4–0.5 mm. long, oblong-ellipsoid. Ovary 2–3 × 1.5 mm., obovoid-cylindric; style up to 6 mm. long, filiform; stigma c. 0.5–0.6 mm. long, linear. Capsule 3–4 × 1–1.5 mm., obovoid-cylindric. Seeds c. 0.5 mm. in diam., subglobose, testa faveolate.

Mozambique. N: Angoche, 18 km. on road to Songage, fl. & fr. 22.i.1968, *Torre & Correia* 17308 (LISC). MS: Cheringoma, near Dondo, *Mendonça* 2482 (LISC). GI: Inhambane, fl. 1.ii.1898, *Schlechter* 12087 (COI; FI; K; P).
Also in Zaire, Angola and S. Africa (Natal). Borders of swamps; c. 20 m.

2. CONGOLANTHUS A. Raynal

Congolanthus A. Raynal in Adansonia, Sér. 2, **8**: 56 (1968).

Annual herbs. Stems erect, slender, often branched, 4-angled. Leaves ovate to elliptic or obovate, linear at the apex of the branches, entire. Flowers 4-merous, axillary, solitary or 2–3-nate. Calyx tube cylindric, with 4 ribs and 8 delicate veins, papyraceous; lobes 1-nerved. Corolla tube cylindric, with 4-lobes. Filaments included in the corolla tube or exserted; anthers narrowly ellipsoid, not twisted. Ovary narrowly ellipsoid, unilocular; ovules numerous; style oblong-cylindric; stigma bilobed. Capsule narrowly-cylindric, bivalved.

A monotypic tropical African genus having affinities with *Neurotheca*.

Congolanthus longidens (N.E. Br.) Raynal in Adansonia, Sér. 2, **8**: 56, t. 2 (1968). —Boutique in Fl. Afr. Centr., Gentianaceae: 17 (1972). TAB. **3**. Syntypes from Nigeria, Zaire and Tanzania.
    Neurotheca longidens N.E. Br. in De Wild. Ann. Mus. Congo, Bot., Sér. 5, **2**: 337 (1908); in De Wild., Pl. Bequaert. **2**: 109 (1923); loc. cit. **4**: 352 (1928). —Baker & N.E. Br. in F.T.A. **4**, 1: 560 (1903). Type as above.
    Neurotheca densa De Wild., Ann. Mus. Congo, Bot., Sér. 5, **2**: 337 (1908). Type from Zaire.

Annual herb 5–50 cm. tall, often branched. Stem 4-angular, with ridges or narrow wings, glabrous. Leaves subsessile; lamina 2–25(35) × 1.3–10(15) mm., ovate, ovate-lanceolate, elliptic to obovate, becoming linear towards the apex of the branches, subacute to obtuse at the apex, base decurrent as a ridge or a narrow wing, 1-nerved, glabrous. Flowers pink whitish, brown punctate, subsessile; pedicel 0–1(3) mm. long. Calyx tube 1–2 mm. long, lanceolate to linear, acute at the apex, membranous and hyaline at the margin. Corolla tube 3–5 mm. long; lobes 1.5–2 mm. long, ovate, subacute at the apex. Filaments 0.5–4 mm. long, slender; anthers 0.5–0.6 mm. long, dorsifixed. Ovary 2.5 mm. long, narrowly ellipsoid; style 1.5–2 mm. long; stigmatic lobes, 0.4–0.6 mm. long, elliptic. Capsule 4 × 0.75–1.1 mm., oblong-cylindric. Seeds 0.5–0.6 mm. in diam., subglobose, brown; testa faveolate.

Tab. 3. CONGOLANTHUS LONGIDENS. 1, flowering branch (×½); 2, portion of flowering branch (×6); 3, flower (×6); 4, corolla opened out (×6); 5, stamen (× 12½); 6, anther (× 25); 7, dorsifixed anther (× 25); 8, pistil (×6); 9, stigmatic lobes (× 25); 10, capsule and accrescent calyx (× 6); 11, seed (× 25), all from *Mutimushi* 3271.

Zambia. W: Mwinilunga, L. Chibesha Rd., fl. & fr. 21.v.1969, *Mutimushi* 3271 (K; NDO).
Also in Nigeria, Central African Republic, Gabon, Zaire, Congo Republic, Uganda and Tanzania.
Swamp forest margins.

3. EXACUM L.

Exacum L., Sp. Pl. **1**: 112 (1753); Gen. Pl., ed 5: 51 (1754).

Annual herbs or rarely perennial. Stems erect simple or branched, 4-ridged, glabrous.
Leaves, linear-lanceolate, narrowly elliptic to ovate or ovate-elliptic, entire, 3–7 nerved.
Flowers 4–5 merous, in teminal or axillary 2-or 3-chotomous cymes. Calyx 4–5 lobed to the
middle or nearly to the base. Corolla regular or somewhat zygomorphic, with a short,
campanulate or broadly cylindrical tube; lobes 4–5, spreading. Stamens 4–5, inserted
below the sinuses; anthers opening by terminal pores prolonged into longitudinal slits
during the dehiscence, not twisted. Ovary subglobose, bilocular; ovules numerous; style
subulate; stigma capitate. Capsule subglobose or obovoid, septicidally bivalved. Seeds
numerous, subglobose-polyhedric; testa faveolate.

A small palaeotropical genus comprising about 65 species mainly from the warm regions of Asia
and also in Tropical Africa, Socotra, Mascarene Islands and Madagascar.

Leaves linear-lanceolate, narrowly elliptic to elliptic, cuneate at the base; corolla 5–8 mm. long in
    fruit; capsule subglobose, 2.5–4 mm. in diam.   -   -   -   -   -   1. *oldenlandioides*
Leaves broadly ovate, rounded to subcuneate at the base; corolla 10–15 mm. long in fruit; capsule
    ellipsoid, 4.5–6 × 3.5–4 mm.   -   -   -   -   -   -   -   -   2. *zombense*

1. **Exacum oldenlandioides** (S. Moore) Klackenb. in Opera Bot. **84**: 88 (1985). Lectotype from
    Zanzibar.
        Sebaea oldenlandioides S. Moore in Journ. Bot. **15**: 68 (1877). Type as above.
        Exacum quinquenervium auct. afr. [Baker & N.E. Br. in F.T.A. **4**, 1: 546 (1903) pro parte excl.
    Mascarene Is. —Hutch. & Dalz., F.W.T.A. **2**, 1: 180 (1931) pro parte excl. Masc. Is. —Taylor in
    F.W.T.A., ed. 2, **2**: 298 (1963) pro parte excl. Madag. —Marais & Verdoorn in Fl. S. Afr. **26**: 211
    (1963) pro parte excl. specim. *Bojer* s.n. —Friedrich-Holzh. in Merxm., Prodr. Fl. SW. Afr. **110**,
    Gentianaceae: 2 (1967). —Boutique in Fl. Afr. Centr., Gentianaceae: 53 (1972) pro parte excl.
    Madag.] non Griseb. (1839).

Annual herb 10–60 m. tall, minutely papillose at the nodes and on the inflorescences
when young, glabrous with age. Stem slender, simple or branched above, 4-ridged. Leaves
petiolate in 4–7 distant pairs; the lower pair smaller; petiole 2–5 mm. long; lamina 10–50 ×
3–10 mm., the upper ones linear-lanceolate, narrowly elliptic to elliptic, acute at the apex,
cuneate at the base, the lower ones smaller and elliptic, rounded at the apex, 3–5 nerved.
Flowers blue, pale violet or rarely white, turning yellow when dry in terminal and axillary
dichotomous cymes; pedicel 4–15 mm. long; bracts 2–10 mm. long, linear, acute. Calyx
tube 0.5–1 mm. long; lobes 2–4.5 × 1–1.5 mm., ovate-elliptic, membranous, hyaline, very
broadly winged on the keel; wings reticulately veined, minutely papillose along the
margin, accrescent in fruit, up to 2.5 mm. broad. Corolla tube 2.5–4 mm. long, broadly
cylindrical, the lower portion enlarged in fruit; lobes erect, 2.25–3 × 1.5–2.5 mm.,
subcircular to ovate or obovate. Stamens inserted 0.5–1 mm. below the corolla sinuses;
filaments 0.75–1.25 mm. long; anthers 0.75–1.75 mm. long. Ovary 2–3 × 1–1.5 mm.,
globose or ovoid-globose; style 1–2 mm. long; stigma capitate, often rather small. Capsule
2.5–4 mm. in diam., globose or obovoid-globose. Seeds 0.25–0.35 mm. in diam.,
subglobose-polyedric, minutely faveolate, brownish.

Zambia. B: Masese, fl. & fr. 14.iii.1961, *Fanshawe* 6438 (K). E: Katete R., near Katete, fl. & fr.
10.iii.1957, *Wright* 178 (K). S: Mazabuka, Kafue Basin Survey, fl. & fr. 29.iii.1963, van *Rensberg* 1856
(K). **Zimbabwe**. N: Gokwe, Sengwa Res. St., fl. & fr. 16.iv.1969, *Jacobsen* 611 (K; SRGH). W: Matobo,
Besna Koblia Farm, fl. & fr. iv.1961, *Miller* 7901 (K; SRGH). S: Gwanda, Doddieburn Ranch, Makoli
Kopje, fl. & fr. 11.v.1972, *Pope* 758 (K; LISC; PRE; SRGH). **Malawi**. S: Blantyre, near B. & C.
Workshop, Ndirande, fl. & fr. *Msinkhu* 22 (K; PRE; SRGH). **Mozambique**. N: Niassa, between
Cuamba and Mutuali, fl. & fr. 24.iv.1961, *Balsinhas & Marrime* 426 (BM; COI; LISC). Z: Zambesia, 32
km. N. of Quelimane, fl. & fr. 20.viii.1962, *Wild* 5879 (COI; K; LISC; PRE; SRGH). MS: Roadside Vila
Pery to Mombane Forest, fl. & fr. 8.iv.1932, *Chase* 4451 (LISC; SRGH).
Tropical Africa from Senegal and southern parts of Chad and Kenya, northeastern Angola to S.
Africa (NE. Transvaal). Sandy pan margins, moist ground and stream and river banks.

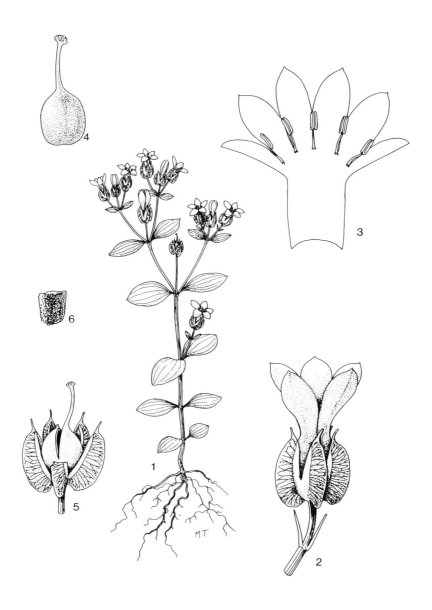

Tab. 4. **EXACUM ZOMBENSE.** 1, habit (× ½); 2, flower (× 3); 3, corolla opened out (× 3); 4, pistil (× 3); 5, dehiscing fruit within calyx (× 3); 6, seed (× 20), all from *Leach & Schelpe* 11488.

2. **Exacum zombense** N.E. Br. in F.T.A. **4**, 1: 546 (1903). TAB. **4**. Lectotype: Malawi, Mount Zomba, *Buchanan* 449 (K).

Annual herb 4–15 cm. tall. Stem glabrous, slender, simple or branched, 4-ridged. Leaves petiolate, in 3–5 distant pairs, the lower pair smaller; petiole 1.5–4 mm. long; lamina 5–30 × 2.5–25 mm., broadly ovate, obtuse or minutely apiculate at the apex, rounded to subcuneate at the base, 5-nerved. Flowers lilac or pale blue, rarely white, drying yellow, in terminal and axillary dichotomous cymes, rarely solitary; pedicel 0.5–20 mm. long; bracts 1.5–10 mm. long, linear to linear-lanceolate, acute. Calyx tube 0.5–1 mm. long; lobes 4–5 × 2.5–3 mm. ovate-elliptic, acutely-cuspidate to abruptly acuminate at the apex, membranous, hyaline, very broadly winged on the keel; wings reticulately veined, minutely papillose along the margin, accrescent in fruit, up to 2 mm. broad. Corolla tube 3.5–6.5 mm. long, broadly cylindrical, the lower portion enlarged in fruit; lobes erect, 4–6 × 2–2.5 mm., oblong-obovate, obtuse at the apex. Stamens inserted 0.5–1 mm. below the corolla sinuses; filaments 1–2 mm. long; anthers 0.75–1 mm. long. Ovary 2–2.5 × 1.5–2 mm., ovoid to subglobose; style 2.5–3 mm. long; stigma capitate. Capsule 4.5–6 × 3.5–4 mm., ellipsoid. Seed 0.3–0.5 mm. diam., subglobose-polyedric, faveolate.

Malawi. S: Zomba Plateau, 1500 m., 5.vi.1946, *Brass* 16249 (K). **Mozambique**. N: Niassa, Amaramba, 20 km. from Cuambo, fl. & fr. 15.ii.1964, *Torre & Paiva* 10609 (LISC). Z: NE. of Namuli Peaks, fl. & fr. 27.vii.1962, *Leach & Schelpe* 11488 (K; LISC). MS: Barue, Choa Mt., 23 km. from Catandica (Vila Gouveia), fl. & fr. 26.iii.1966, *Torre & Correia* 15427 (LISC).
Known only from Malawi and N. of Mozambique. Moist grassy banks, wet flush on rocks in grasslands; c. 1100 m.

4. SEBAEA R. Br.

Sebaea R. Br., Prodr.: 451 (1810).
Parasia Rafin., Fl. Tellur. **3**: 78 (1836).
Lagenias E. Mey., Comm. Pl. S. Afr.: 186 (1837).
Belmontia E. Mey., tom. cit.: 183 (1837).
Exochaenium Griseb. in DC., Prodr. **9**: 85 (1845).

Annual or perennial (rarely parasitic) erect, procumbent or dwarf herbs. Stems simple or branched, terete to somewhat tetragonous, usually more or less 4-ridged or winged. Leaves well-developed or reduced and scale-like, petiolate or sessile. Flowers 4–5-merous, terminal, solitary or in cymes. Calyx 4–5-lobed, almost free or forming a short tube; lobes lanceolate to ovate or obovate, keeled or winged, papillose or scabrid on the keels. Corolla tube cylindric or funnel-shaped, with 4–5(6) lobes. Stamens 4–5(6); filaments filiform inserted at the sinuses of the corolla or in the tube; anthers oblong, exserted or included, without or with 1 apical gland, without or with 1–2 basal glands. Ovary ovoid, obovoid or almost globose, bilocular; ovules numerous; style filiform, long or short, exserted or included, sometimes with middle or basal swelling; stigma capitate, clavate more or less bilobed. Capsule ovoid or obovoid, membranous or coriaceous, septicidally bivalved. Seeds minute, very numerous, cubical, ridged or frilled.

A genus with c. 159 species widespread throughout tropical Africa and Madagascar. One species extending to India and China, and two species are found in Australia and New Zealand.

1. Flowers solitary or in lax few-flowered cymes; leaf lamina linear, sometimes oblong-lanceolate or rarely triangular-lanceolate, oblong-ovate, usually 0.5–4 mm., up to 6 mm. broad (when up to 10 mm., then leaf lamina linear to oblong-ovate)   -   -   -   -   -   -   2
   - Flowers in dense, rarely lax many-flowered cymes; leaf lamina circular, subreniform or ovate-oblong to elliptic, 7–20 mm. broad   -   -   -   -   -   -   -   -   18
2. Flowers small; corolla tube 3–10 (12) mm. long; capsule 1.5–5.5 mm. long   -   -   -   3
   - Flowers large; corolla tube (10) 12–30 mm. long; capsule 6–10 mm. long   -   -   -   16
3. Stamens exserted; filaments inserted in the corolla sinuses   -   -   -   -   -   4
   - Stamens included; filaments inserted at the middle of the corolla tube or below   -   7
4. Anthers with an apical gland   -   -   -   -   -   -   -   -   -   -   -   5
   - Anthers without an apical gland   -   -   -   -   -   -   -   -   -   -   6
5. Stems below sometimes glandular or glandular-denticulate; calyx lobes 5–7 (9) mm. long; corolla lobes 5–6 (8) mm. long; anthers 1.25–1.5 (2.5) mm. long   -   -   -   -   -   *2. bojeri*
   - Stems not glandular; calyx lobes 3–4 (5) mm.; corolla lobes 2–4.5 (6);
   anthers 0.75–1.25 mm. long   -   -   -   -   -   -   -   -   -   -   *3. junodii*

6. Stems below sometimes glandular or glandular-denticulate; calyx lobes winged on the keel [wing 0.4–0.5 (1.75) mm. broad at the base] - - - - - - - - - - - *2. bojeri*
- Stems not glandular; calyx lobes narrowly winged on the keel (wing c. 0.1 mm. broad) - - - - - - - - - - - - - - - *1. microphylla*
7. Slender herbs up to 20 cm. tall; leaf lamina oblong-ovate, ovate-elliptic, oblong-lanceolate to linear-lanceolate - - - - - - - - - - - - - - - - - 8
- Erect herbs up to 50 cm. tall; leaf lamina linear to oblong-lanceolate - - - - 13
8. Dwarf herbs; stem winged - - - - - - - - - - - *10. perpusilla*
- Erect herbs; stem ridged, not winged - - - - - - - - - - 9
9. Flowers long-pedicellate; pedicels 20–70 mm. long - - - - - - 10
- Flowers short-pedicellate; pedicels 2–15 mm. long - - - - - - 11
10. Leaves (middle of stem) broadly elliptic, 5–6 × 3–4 mm.; calyx segments 4–6 mm. long, dorsally winged, wing wider at the middle, 0.25–0.3 mm. broad; capsule ellipsoid, 4–5.5 × 3–3.5 mm. - - - - - - - - - - - - - *8. fernandesiana*
- Leaves (middle of stem) linear-lanceolate, 1.5–10 × 0.9 mm.; calyx segments 6–8 mm. long, winged, with the wing wider at the base (up to 1.5 mm. broad); capsule subglobose, 2.5 × 2 mm. - - - - - - - - - - - - - - - - *11. alata*
11. Herbs up to 20 cm. tall; wing of the calyx lobes 0.7–1.2 mm. broad; anthers with a short apical gland, 0.1–0.15 mm. long - - - - - - - - - - *12. platyptera*
- Herbs up to 6 cm. tall; wing of the calyx lobes 0.25–0.5 mm. broad; anthers with a long apical gland, 0.25–1.5 mm. long - - - - - - - - - - - - - 12
12. Calyx segments 3–4 mm. long, keeled or dorsally narrowly winged, wing up to 0.25–0.3 mm. broad; apical gland of the anther up to 0.5 mm. long - - - - - - *4. pumila*
- Calyx segments 4.5–6 mm. long, dorsally winged, wing up to 0.5 mm. broad; apical gland of the anther up to 1.5 mm. long - - - - - - - - - *9. africana*
13. Anthers clearly caudate at the base - - - - - - - - - *7. caudata*
- Anthers not caudate at the base - - - - - - - - - - - 14
14. Leaves linear to linear-lanceolate up to 3.5 × 1.5 mm.; corolla tube up to 5 mm. long; apical glands of the anthers 0.1–0.2 mm. long - - - - - - - - - *13. baumiana*
- Leaves linear to linear-lanceolate up to 22 × 2 mm.; corolla tube up to 9 mm. long; apical glands of the anthers 0.5–0.8 mm. long - - - - - - - - - - - 15
15. Calyx lobes equal, 3.5–6 mm. long; anthers with basal, subglobose glands - *6. minuta*
- Calyx lobes unequal, 6–9 mm. long; anthers without basal glands - - - *5. gracilis*
16. Leaves linear to linear-lanceolate, 15–70 × 1–2 mm.; corolla lobes 4–5 mm. long - - - - - - - - - - - - - *16. clavata*
- Leaves linear-lanceolate, oblong-lanceolate to ovate-lanceolate, 5–40 × 2–10 mm.; corolla lobes 5–25 mm. long - - - - - - - - - - - - - - - - 17
17. Wing of the calyx lobes narrow (0.5–0.6 mm. broad) and wider near the middle; anthers with a short apical gland, 0.4–0.6 mm. long - - - - - - - - *14. teuszii*
- Wing of the calyx lobes 1–2 mm. broad and wider near the base; anthers with a long apical gland, 0.75–1.5 mm. long - - - - - - - - - - - *15. grandis*
18. Leaves crowded near the base, the lowest subrosulate narrowed into a petiole-like base; the upper sessile in 3–5 distant pairs; lamina up to 4 cm. long, ovate-oblong to elliptic, rounded or cuneate at the base - - - - - - *17. pentandra* var. *burchellii*
- Leaves sessile, close towards the base but never rosulate, in many pairs, scattered along the stems; lamina up to 2.5 cm. long, elliptic-lanceolate to ovate-subcircular or subcircular-reniform, cordate to rounded-subcordate at the base - - - - - - - - - - 19
19. Corolla up to 7.5 mm. long; corolla tube 3–4 mm. long; lobes 1.5–3.5 × 0.8–1.2 mm. - - - - - - - - - - *18. brachyphylla*
- Corolla longer than 7.5 mm. (up to 20 mm.); corolla tube more than (2.5) 4.5 mm. long; lobes 4.5–11 × 2.5–5.5 mm. - - - - - - - - - - - - - - - 20
20. Inflorescence of lax or compact cymes, terminal and solitary or somewhat corymbose; calyx lobes up to 11 × 5.5 mm.; corolla lobes up to 11 × 5.5 mm.; filaments up to 4 mm. long; anthers 2–4.25 mm. long - - - - - - - - - - - *20. longicaulis*
- Inflorescence corymbose or paniculate-corymbose subdense or dense cymes, sometimes much contracted and head-like; calyx lobes up to 8 × 3 mm.; corolla lobes up to 7.25 × 3.5 mm.; filaments up to 2 mm.; anthers 1–2.5 mm. long - - - - - - - - 21
21. Perennial herbs; taproot well developed with annual, simple or branched erect stems; inflorescene a dense corymb, sometimes much contracted and head-like - - - - - - - - - - *21. sedoides* var. *confertiflora*
- Annual or perennial herbs; without a taproot; inflorescence a subdense corymb or paniculate-corymb, sometimes contracted and almost head-like - - - - - *19. leiostyla*

1. **Sebaea microphylla** (Edgew.) Knobl. in Bot. Zentralbl. **60**: 324 (1894). —Boutique in Fl. Afr. Centr., Gentianaceae: 50 (1972). Type from Angola.
  Cicendia microphylla Edgew. in Trans Linn. Soc. **20**: 83 (1851). Type as above.
  Sebaea welwitschii Schinz, Viertelj. Nat. Gesellsch. Zürich. **36**: 321 (1891). —Hiern, Cat. Pl. Afr. Welw. **1**, 3: 705 (1898). —Baker & N.E. Br. in F.T.A. **4**, 1: 550 (1903). Type from Angola.

Erect annual herb (sometimes parasitic) 4–40 cm. tall. Stem slender, filiform, unbranched or branched above, 4-ridged. Leaves 2–6.5 × 0.5–1.5 mm., oblong-ovate to triangular-lanceolate, acute at the apex. Flowers pale yellow, yellow to gold-yellow, terminal, solitary or in few to many-flowered cymes, pedicellate; pedicels 0.5–13 mm. long. Calyx almost without tube; lobes 5, 3–7 mm. long, linear-lanceolate, subulate-lanceolate to ovate-lanceolate, narrowly winged on the keel (wing c. 0.1 mm. broad). Corolla tube 4–6 mm. long, cylindrical or funnel-shaped, narrowed around the ovary, enlarged from the point of the insertion of the filaments; lobes ovate-elliptic to oblong, 3–6 mm. long, attenuate-acute and sometimes apiculate at the apex. Stamens inserted near the throat of the corolla; filaments 1.5–2 mm. long, reaching the middle of the corolla lobes; anthers 0.7–2 mm. long, oblong, ellipsoid, without apical gland. Ovary oblong-ellipsoid, 2–3.5 mm. long; style 2–3 mm. long; stigma 0.5–1.2 mm. long, globose-ellipsoid, to oblong-ellipsoid, bilobed. Capsule oblong-ellipsoid 3–5 mm. long. Seeds 0.5–0.15 mm. in diam., cubical; testa faveolate.

Zambia. C: Chakwenga Headwaters, 100–129 km. E. of Lusaka, fl. & fr. 27.iii.1965, *Robinson* 6465 (K). S: 12 km. N. of Choma, fl. 23.iii.1957, *Robinson* 2171 (K). **Zimbabwe**. N: Urungwe (Hurungwe) Nat. Park, on Harare to Chirundu Rd., fl. & fr. 15.ii.1981, *Philcox, Leppard & Dini* 8596 (K). S: Bikita, fl. & fr. 10.v.1969, *Biegel* 3088 (K; PRE; SRGH).

Also in Zaire, Rwanda, Burundi, Angola and Tanzania. In open woodland, grassland and savanna; 780–1200 m.

2. **Sebaea bojeri** Griseb., Gen. Spec. Gentianaceae: 169 (1839). —Schinz in Mitt. Geogr. Ges. Lübeck **17**: 30 (1903). —Marais & Verdoorn in Fl. S. Afr. **26**: 1931, fig. 26, 1 (1963). —Ross in Bot. Surv. Mem. **39**: 278 (1972). Type from Madagascar.

Sebaea mirabilis Gilg in Engl., Bot. Jahrb. **26**: 92 (1898). —Schinz in Mitt. Geogr. Ges. Lübeck **17**: 35 (1903); in Bull. Herb. Boiss., Sér. 2, **6**: 726 (1906). Type from S. Africa (Transvaal).

Sebaea pratensis Gilg in Engl., Bot. Jahrb. **30**: 377 (1901). —Schinz in Mitt. Geogr. Ges. Lübeck **17**: 38 (1903). —Baker & N.E. Br. in F.T.A. **4**, 1: 550 (1903). Type from Tanzania.

Sebaea saccata Schinz. in Mitt. Geogr. Ges. Lübeck **17**: 25 (1903); in Bull. Herb. Boiss., Sér. 2, **6**: 726 (1906). —Hill & Prain in F.C. **4**, 1: 1072 (1909). Type from S. Africa. (Transvaal).

Erect annual herb 5–20 cm. tall. Stem slender unbranched or branched from the base, sometimes glandular or glandular denticulate in the lower part. Leaves 4–6 (10) × 1 (3) mm., linear-lanceolate to ovate-lanceolate, acute or acuminate at the apex, the lower ones reduced and scale-like. Flowers yellow, solitary or in few to many-flowered cymes, pedicellate; pedicels 1–1.5 mm. long. Calyx with a very short tube; lobes 5, 5–7 (9) × 1–2.25 mm., ovate-lanceolate to lanceolate, acuminate at the apex, winged on the keel, with hyaline margins; wing 0.4–0.5 (0.75) mm. broad near the base of the calyx. Corolla tube 3–5 mm. long; lobes 5–6 (8) × 2.25–3 (4) mm. Stamens 1.3–2.75 mm. long, inserted in the corolla-sinuses; anthers 1.25–1.5 (2.5) mm. with or without a small apical, sessile gland. Ovary 2.5 × 1.5 mm., ovoid; style 3.5 mm. long; stigma bilobed with broad lobes, 0.9–1 mm. long. Capsule (immature) 3 × 2 mm., ovoid.

Zimbabwe. E: Chimanimani Mts., fl. & fr. imat. 11.iv.1967, *Grosvenor* 337 (K). **Malawi**. N: Nyika Plateau, Chelinda Camp, fl. & fr. immat. 18.iv.1970, *Brummitt* 10874 (K).

Also in S. Africa and Tanzania. In open woodland; 1500–2300 m.

3. **Sebaea junodii** Schinz in Bull. Herb. Boiss., Sér. 2, **4**: 442 (1896). —Hill & Prain in F.C. **4**, 1: 1069 (1909). —Marais & Verdoorn in Fl. S. Afr. **26**: 1931, fig. 26, 4 (1963). —Boutique in Fl. Afr. Centr., Gentianaceae: 50 (1972). —Ross in Bot. Surv. Mem. **39**: 278 (1972). TAB. **8**. Type from Natal.

Erect annual herb 3.5–20 cm. tall. Stem slender, unbranched or 2–3 branched. Leaves 1–2.5 (5) × 0.5–1.8 mm., lanceolate to narrowly-ovate, acute at the apex, the lower ones reduced and scale-like. Flowers whitish or lemon yellow to yellow, solitary or in few-flowered cymes, pedicellate; pedicels 1–5 mm. long. Calyx with a very short tube; lobes 5, 3–4 (5) mm. long, ovate-lanceolate to lanceolate, acuminate to mucronate at the apex, winged on the keel, (wing up to 0.4–0.5 mm. broad), with hyaline margins; corolla tube 3–4 (5) mm. long, cylindric or funnel-shaped; lobes 5, 2–4.5 (6) mm. long, ovate-lanceolate to oblong-ovate, subacute or rounded at the apex. Stamens inserted in the corolla-sinuses; filaments 0.8–1 (2) mm. long; anthers 0.75–1.25 mm. long, linear ellipsoid, with a narrowly clavate apical gland c. 0.12–0.17 mm. long. Ovary ellipsoid, 1–1.5 mm. long; style 3–4 mm. long; stigma 0.7–1 mm., oblong bilobed. Capsule 2–2.5 mm., ellipsoid. Seeds 0.05–0.1 mm. in diam., cubical or subglobose; testa faveolate.

Zambia. N: Mwenda Hills, fl. & fr. 5.v.1957, *Richards* 9579 (K). **Zimbabwe**. C: Wedza Mt. near Schelite mine, fl. & fr. 7.iv.1964, *Wild* 6516 (K; LISC; PRE; SRGH). E: Mutare, fl. & fr. 23.iii.1958,

Chase 6861 (K; SRGH). **Mozambique**. M: Maputo, Libombos, near Namaacha, Mponduine Mt., fl. 22.ii.1955, *Exell, Mendonça & Wild* 510 (LISC; SRGH).

Also in Zaire, S. Africa and Tanzania. Soils with rich leaf mould; 800 m.

Exell, Mendonça & Wild 510 is a more delicate plant with a very lax inflorescence and with the pedicels spreading. It may represent a different taxon.

4. **Sebaea pumila** (Baker) Schinz in Bull. Herb. Boiss., Sér. 2, **6**: 731 (1906). —Taylor in F.W.T.A. ed. 2, **2**: 299 (1963). —Boutique in Fl. Afr. Centr., Gentianaceae: 47 (1972). Type from Nigeria.

Belmontia pumila Baker in Kew Bull. **1894**: 25 (1894). —Baker & N.E. Br. in F.T.A. **4**, 1: 552 (1903). —Hutch. & Dalz. F.W.T.A. **2**, 1: 183 (1931). Type as above.

Exochaenium pumilum (Baker) Hill in Kew Bull. **1908**: 336 (1908). Type as above.

Small annual, erect herbs, up to 6 cm. tall. Stem unbranched, or branched, 4-ridged. Leaves 3–8 × 1–3 mm., oblong-lanceolate to oblong ovate, subacute at the apex, spreading, sometimes the lower deflexed. Flowers white with a yellow centre at the base of the corolla tube, solitary, terminal or axillary, pedicellate; pedicels 2–10 mm. long. Calyx 5 lobed, almost without tube; lobes 3–4 mm. long, ovate-lanceolate to oblong-lanceolate, acute at the apex, with a narrowly-winged green keel (wing up to 0.25–0.3 mm. broad). Corolla tube 3–4.5 mm. long, narrowly funnel-shaped in the upper half; lobes 5, obovate, 1–2 mm. long. Stamens inserted below the middle of the corolla tube; anthers 0.6–0.7 mm. long, with a narrowly clavate apical gland, 0.25–0.5 mm. long, and with 2 subglobose glands at the base. Ovary subglobose, 1.5 mm. in diam.; style 1–1.2 mm. long; stigma, oblong, 0.5–0.6 mm. long. Capsule subglobose, 2–3 mm. in diam. Seeds 0.1–0.2 mm. in diam., cubical; testa faveolate.

Zambia. N: Mbala Distr., Lumi R., fl. & fr. 30.iii.1957, *Richards* 8939 (K). W: Matonchi, Mwinilunga, fl. & fr. 16.iv.1960, *Robinson* 3622 (K). **Mozambique**. Z: Quelimane Distr., Mocuba, fl. & fr. 18.v.1948, *Faulkner* 265 (K).

Also in Nigeria, Zaire, Uganda and Tanzania. In damp mud, in marshes or swampy ground among fairly long grass; 1400–1680 m.

5. **Sebaea gracilis** (Welw.) Paiva & Nogueira in Anal. Jard. Bot. Madrid **47**, 1: xxx (1990). Type from Angola.

Belmontia gracilis Welw. in Trans. Linn. Soc. **27**: 47 (1869). —Baker & N.E. Br. in F.T.A. **4**, 1: 551 (1903). Type as above.

Parasia gracilis (Welw.) Hiern, Cat. Afr. Pl. Welw. **1**, 3: 708 (1898). Type as above.

Exochaenium gracilis (Welw.) Schinz in Bull. Herb. Boiss., Sér. 2, **6**: 716 (1906). Type as above.

Erect, delicate, annual herb up to 50 cm. tall. Stem-slender, unbranched or very sparingly branched, 4-ridged. Leaves 8–22 × 1.5–2 mm., linear-lanceolate. Flowers pale cream, pale yellow, terminal, solitary, pedicellate; pedicels up to 0.5 mm. long. Calyx with 5 unequal segments, 6–9 mm. long, lanceolate, acuminate, with a narrowly winged keel. Corolla tube 7–8.5 mm. long, cylindric, widening at level of the stamens, externally pubescent-verrucous; lobes 5, 2–3 mm. long, broadly ovate. Stamens included, inserted at the midlle of the corolla tube; filaments short, c. 1–1.2 mm. long, subulate; anthers oblong-ellipsoid, with a narrowly clavate and papillate apical gland, c. 0.5 mm. long, without basal glands. Ovary obovoid, 4 × 2 mm.; style c. 1 mm. long; stigma bifid, papillose. Capsule obovoid 4–5 × 2–3 mm. Seeds c. 0.1–0.2 mm. in diam., numerous, cubical; testa faveolate.

Zambia. W: Kitwe, fl. & fr. 26.iv.1957, *Fanshawe* 3197 (K).

Also in Angola. In dambo; c. 1600 m.

6. **Sebaea minuta** Paiva & Nogueira in Anal. Jard. Bot. Madrid **47**, 1: xxx (1990). Type: Zimbabwe, Bulawayo, *Eyles & Johnson* 1032 (K, holotype).

Exochaenium exiguum Hill in Kew Bull. **1909**: 50 (1909). Type as above.

Annual erect herb up to 30 cm. tall. Stem unbranched or sparingly branched above, 4-ridged. Leaves 2–12 × 1–2 mm., linear-lanceolate, acute at the apex, attenuate at the base. Flowers cream, yellow or white, solitary or in a few-flowered cyme, pedicellate; pedicels up to 8 cm. long. Calyx with 5 segments, 3.5–6 mm. long, lanceolate, acute or caudate at the apex with a very narrowly winged keel. Corolla tube up to 9 mm. long, cylindric or funnel-shaped; lobes 2.5 × 1–2.25 mm., obovate, rounded and apiculate at the apex; stamens included, inserted in the corolla tube near the base; filaments 1–1.5 mm. long; anthers oblong-ellipsoid, c. 0.75 mm. long, with a narrowly clavate apical gland,

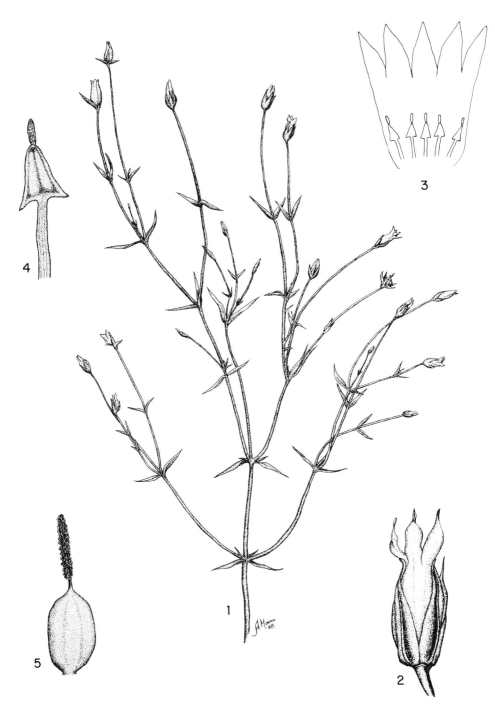

Tab. 5. SEBAEA CAUDATA. 1, flowering stem (× 1¾); 2, flower (× 4½); 3, corolla opened out showing stamens (× 4); 4, stamen (× 19); 5, pistil (× 7½), all from *Robinson* 1251.

0.5–0.8 mm. long, and with 2 minute subglobose, stipitate glands at the base. Ovary 3 × 2 mm., ellipsoid; style c. 2 mm. long; stigma 1.5–1.75 mm. long, clavate, papillate. Capsule 4–5 × 3–4 mm., ellipsoid. Seeds 0.2–0.25 mm. in diam., cubical; testa faveolate.

Zambia. N: Chinsali Distr., Shiwa Ngandu, Lake Young, fl. & fr. 17.i.1959, *Richards* 10712 (K). W: 7 km. from Chizela, fl. & fr. 27.iii.1961, *Drummond & Rutherford-Smith* 7420 (K; LISC; PRE). C: Chiwefwe, fl. & fr. 1.v.1957, *Fanshawe* 3238 (K). S: Simasunda dambo, 3 km. E. of Mapanza, fl. & fr. 28.iii.1954, *Robinson* 642 (K). **Zimbabwe**. N: Gokwe, fl. & fr. 16.iv.1968, *Jacobsen* 618 (K). W: Matobo Distr., Farm Chasterfield, fl. & fr. iii.1958, *Miller* 5129 (K; PRE). C: Harare, Hatfield, fl. & fr. 7.iv.1957, *Whellan* 1223 (K; PRE). S: Bikita, fl. & fr. 5.v.1969, *Biegel* 3008 (K). **Mozambique**. N: Malema, Inago, fl. & fr 20.iii.1964, *Torre & Paiva* 11286 (LISC).
Also in Angola. Growing at edge of laterite pan in marshy soils; 850–1500 m.

7. **Sebaea caudata** Paiva & Nogueira in Anal. Jard. Bot. Madrid **47**, 1: xxx (1990). TAB. **5**. Type: Zambia, Mapanza, *Robinson* 1251 (K, holotype).

Erect annual herb up to 30 cm. long. Stem slender, sparingly branched, 4-ridged. Leaves 10–14 × 1–1.5 mm., linear-lanceolate, acute at the apex, attenuate at the base. Flowers cream or pale yellow, in few-flowered cymes, pedicellate; pedicels 2.5–6 cm. long. Calyx segments 5, 4.5–5.5 mm. long, lanceolate, caudate at the apex, with the keel narrowly winged. Corolla tube as long as the calyx, funnel-shaped; lobes 3 mm. long, broadly ovate, apiculate. Stamens inserted in the corolla tube near the base; filaments 1–1.2 mm. long, filiform; anthers 0.8–0.9 × 0.3–0.4 mm., oblong-ellipsoid, with a linear clavate apical gland, 0.35–0.4 mm. long, caudate at the base. Ovary c. 3 × 2 mm. long, ellipsoid; style c. 0.7–0.8 mm. long; stigma c. 1.5 mm. long, oblong-linear, papillose, exserted. Capsule 3.5–4 × 2–2.5 mm., ellipsoid. Seeds 0.2–0.25 mm. in diam., cubical; testa faveolate.

Zambia. S: Mapanza, Simansunda Dambo, fl. & fr. 27.iv.1955, *Robinson* 1251 (K).
Known only from Zambia. Dambo.

8. **Sebaea fernandesiana** Paiva & Nogueria in Anal. Jard. Bot. Madrid **47**, 1: xxx (1990). TAB. **6**. Type: Zambia, Mwinilunga, Zambezi Rapids, c. 6 km. N. of Kalene Hill, *Hooper & Townsend* 256 (K, holotype).

Erect annual herb up to 15 cm. tall. Stem slender, unbranched or branched from the base, 4-ridged. Leaves 3–8 × 2–4 mm., ovate-elliptic acute at the apex, attenuate at the base into a short petiole. Flowers pale creamy to white, solitary, terminal; puduncles 2–7 cm. long. Calyx segments 5, 4–6 mm. long, lanceolate, acute to acuminate at the apex, with a dorsal wing up to 0.25–0.3 mm. broad at the middle of the segments. Corolla tube 9–10 mm. long, lobes 3–3.5 mm. long, obovate, apiculate at the apex. Stamens inserted in the corolla tube near the base; filaments c. 2 mm. long; anthers, oblong-ellipsoid, 0.7–1 mm. long, with a 1–1.5 mm. long, narrowly clavate apical gland, and two minute globose, stipitate, basal glands. Ovary c. 2 mm. in diam.; style c. 1.8 mm. long; stigma c. 1 mm. long, clavate, papillate. Capsule 4–5.5 × 3–3.5 mm., ellipsoid. Seeds 0.2–0.25 mm. in diam., cubical; testa faveolate.

Zambia. W: Mwinilunga Distr., Zambezi Rapids, c. 6 km. N. of Kalene Hill, fl. & fr. 26.ii.1975, *Hooper & Townsend* 256 (K).
Known only from western Zambia. In damp soil over rocky outcrops; 1350 m.

9. **Sebaea africana** Paiva & Nogueira in Anal. Jard. Bot. Madrid **47**, 1: xxx (1990). TAB. **7**. Type: Zambia, Kawambwa-Mbereshi Rd., *Richards* 9337 (K, holotype).

Erect annual herb up to 6 cm. tall. Stem unbranched above, 4-ridged. Leaves 3–8 × 0.5–1 mm., linear-lanceolate, the lowermost smaller. Flowers with the tube pale yellow and the lobes white, solitary, terminal, pedicellate; pedicels 5–15 mm. long. Calyx segments 5, 4.5–6 mm. long, lanceolate, acuminate with a dorsal wing up to 0.4–0.5 mm. broad near the base, narrowing towards the apex. Corolla tube up to 9 mm. long, funnel-shaped; lobes 3–4 mm. long, ovate-oblong. Stamens included, inserted near the base of the tube; filaments 0.5–0.6 mm. long; anthers oblong-ellipsoid, c. 0.9 mm. long, with a narrowly clavate apical gland, 1–1.5 mm. long and two minute globose stipitate basal glands. Ovary 1.1–1.2 mm. in diam., subglobose; style 1.5 mm. long; stigma 1.5 mm. long,

16

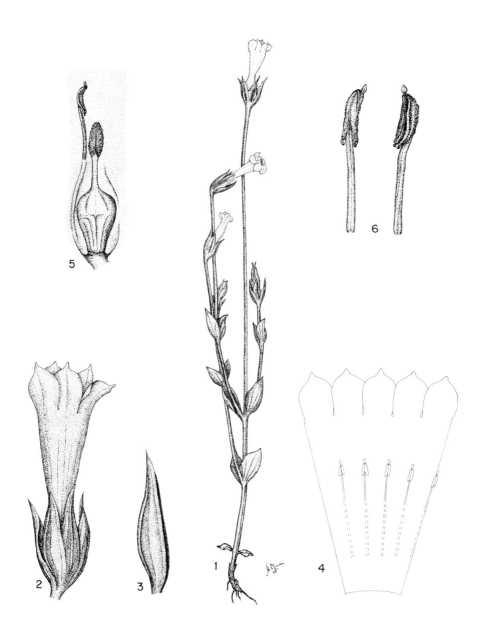

Tab. 6. SEBAEA FERNANDESIANA. 1, habit (× 1½); 2, flower (× 6); 3, calyx lobe (× 7⅓); 4, corolla opened out to show stamens (× 7); 5, part of corolla showing pistil and one stamen (× 7½); 6, stamen, two aspects (× 15⅛), all from *Hooper & Townsend* 256.

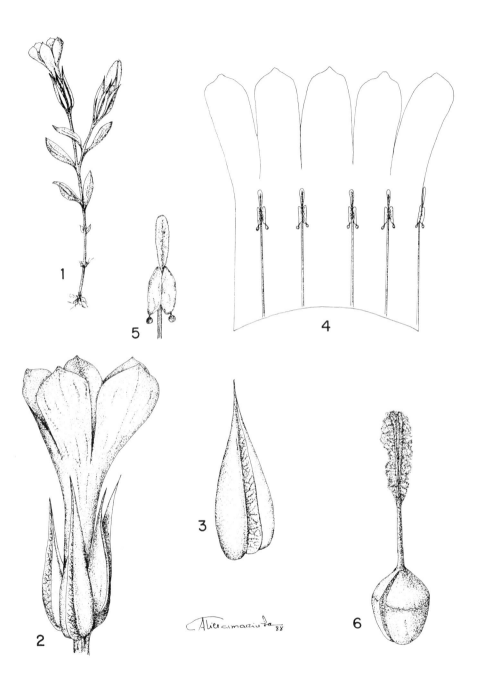

Tab. 7. SEBAEA AFRICANA. 1, habit (× 1½); 2, flower (× 6); 3, calyx lobe, showing dorsal wing (× 12½); 4, corolla opened out to show stamens (× 6); 5, anther (× 12½); 6, pistil (× 12½), all from *Richards* 9337.

clavate, papillate. Capsule 3 × 2 mm., ellipsoid. Seeds 0.2–0.3 mm. in diam., cubical; testa faveolate.

Zambia. N: Kawambwa-Mbereshi Rd., fl. & fr. 19.iv.1957, *Richards* 9337 (K).
Not known elsewhere. In damp sandy ground among grass; 1680 m.

10. **Sebaea perpusilla** Paiva & Nogueira in Anal. Jard. Bot. Madrid **47**, 1: xxx (1990). Type: Zambia, Mwinilunga Distr., Sinkabolo Dambo, fl. 9.xii.1937, *Milne-Redhead* 3579 (K, holotype).
    Exochaenium pygmaeum Milne-Redhead in Kew Bull. **4**: 337 (1951). Type as above.

Dwarf annual or, perhaps, perennial with fleshy roots, up to 5.5 cm. tall. Stem erect, branched from the base, winged. Leaves 4 × 3.5 mm., broadly ovate, rounded at the base, acute at the apex. Flowers white, with a yellow tube, in few to many-flowered cymes, pedicellate; pedicels 0.3–1.3 mm. long. Calyx segments 5–6, c. 4 mm. long, lanceolate, acute or acuminate at the apex, broadly winged on the keel; keel widening near the base. Corolla tube 3.5–4 mm. long, narrowly funnel-shaped in the upper half; lobes 5–6, 2.5–3 mm. long, obovate, stamens inserted below the middle of the corolla tube; filaments 1–3 mm. long; anthers c. 0.6 mm. long, oblong-ellipsoid, with a narrowly clavate and papillate apical gland up to 6 mm. long, and 2 minute subglobose, stipitate basal glands. Ovary c. 1–2 mm. long, subglobose; style c. 0.4 mm. long; stigma 2 mm. long, oblong, papillose, bilobed. Capsule c. 3 × 2.5 mm., subglobose. Seeds 0.1–0.2 mm. in diam., cubical; testa faveolate.

Zambia. B: Mongu, fl. & fr. 29.i.1966, *Robinson* 6829 (K). W: 37 km. W. of Mwinilunga, Matonchi Rd., fl. & fr. 24.i.1975, *Brummitt, Chisumpa & Polhill* 14093 (K).
Known only from Zambia. Wet dambo.

11. **Sebaea alata** Paiva & Nogueira in Anal. Jard. Bot. Madrid **47**, 1: xxx (1990). Type: Zambia, Chishinga Ranch near Luwingu, *Astle* 545 (K, holotype).

Erect annual herb up to 15 cm. tall. Stem slender, unbranched or branched above, 4-ridged. Leaves 1.5–10 × 0.9 mm., linear-lanceolate, the lowest smaller. Flowers creamy pale yellow, terminal, solitary, pedunculate; peduncle (1.5) 2 cm. long. Calyx segments 5, 6–8 mm. long, lanceolate, acuminate, with a dorsal wing up to 1.5 mm. broad near the base, narrowing towards the apex. Corolla tube 12 mm. long, funnel-shaped; lobes 3.5–4 mm. long, ovate-oblong, rounded at the apex. Stamens included, inserted near the base of the corolla tube; filaments very short; anthers c. 1.2 mm. long, ellipsoid with a narrowly clavate, apical, c. 1 mm. long gland and two minute globose stipitate basal glands. Ovary c. 1.5 mm. in diam., subglobose; style c. 2 mm. long; stigma c. 2.5 mm. in diam. Capsule 2.5 × 2 mm., subglobose. Seeds 0.2–0.25 mm. long, cubical; testa faveolate.

Zambia. N: Chishinga Ranch near Luwingu, fl. & fr. 27.iv.1961, *Astle* 545 (K).
Known only from northern Zambia. In dambo, widespread but not very common; 1580 m.

12. **Sebaea platyptera** (Baker) Boutique in Fl. Afr. Centr., Gentianaceae: 48 (1972). Type from Angola.
    Belmontia platyptera Baker in Kew Bull. **1894**: 25 (1894). —Baker & N.E. Br. in F.T.A. **4**, 1: 552 (1903). Type as above.
    Exochaenium platypterum (Baker) Schinz in Bull. Herb., Boiss. Sér. 2, **6**: 716 (1896). Type as above.
    Parasia platyptera (Baker) Hiern, Cat. Pl. Afr. Welw. **1**, 3: 706 (1898). Type as above.

Erect annual herb up to 20 cm. tall. Stem simple or branched above, 4-ridged. Leaves scattered, 3–18 × 1.5–8 mm., oblong-ovate, acute or obtuse at the apex, cuneate at the base. Flowers white, pale yellow, solitary, terminal or axillary, pedicellate; pedicels up to 12 mm. long. Calyx segments 5, 4–8 mm. long, ovate to lanceolate, acuminate at the apex, broadly winged on the keel; wing 0.7–1.2 mm. broad. Corolla tube 7–9 mm. long, cylindric, enlarged at the point of insertion of the stamens; lobes 5, c. 2 mm. long, obovate; stamens included, inserted at the middle of the corolla tube; filaments c. 0.2 mm. long; anthers 0.8 mm. long, with a narrowly clavate apical, 0.1–0.15 mm. long, gland. Ovary 3.5–4 mm., narrowly ovoid; style 1–1.2 mm. long; stigma 0.5–0.6 mm. long, oblong. Capsule 3–4 × 2.3 mm., ovoid. Seeds 0.2–0.25 mm. in diam., cubical; testa faveolate.

Zambia. N: Mbala Distr., Kali Dambo, near Kawimbe Mission, fl. & fr. 6.v.1952, *Richards* 1615 (K). S: Mazabuka, fl. & fr. 20.iii.1958, *Robinson* 2812 (K; SRGH). **Zimbabwe**. N: Mwami (Miami), fl & fr. iv.1926, *Rand* 67 (BM). C: Harare, fl. & fr. 26.iv.1953, *Greatrex* in GHS.42604 (K; LISC). S: Masvingo, fl. & fr. 1909–12, *Monro* 1775 (BM). **Malawi**. S: Ntcheu Distr., Lower Kirk Range, Chipusiri, fl. & fr. 17.iii.1955, *Exell, Mendonça & Wild* 935 (BM; LISC).
Also in Zaire, Burundi, Angola and Tanzania. Wet grassland, in sandy places; 1460–1600 m.

13. **Sebaea baumiana** (Gilg) Boutique in Fl. Afr. Centr., Gentianaceae: 46 (1972). Type from Angola.
Belmontia baumiana Gilg in Baum, Kunene-Samb.-Exped. 331 (1903). —Baker & N.E. Br. in F.T.A. **4**: 626 (1904). Type as above.

Erect annual herb up to 35 cm. tall. Stems slender, filiform, unbranched or rarely branched, 4-ridged. Leaves scattered, 3.5 × 0.7–1.5 mm., linear or linear-lanceolate, acute at the apex. Flowers white, ivory, lemon-yellow to primrose-yellow, solitary, terminal, pedicellate; pedicels 3–15 cm. long. Calyx segments 5, 3–4 mm. long, lanceolate, acuminate. Corolla tube up to 5 mm. long, cylindric, inflated at level of anthers and puberulent, verrucose externally; lobes 1.7 (3) mm. long, broadly ovate, acute. Stamens included, inserted at the middle of the corolla tube; filaments 0.2–0.3 mm. long; anthers narrowly ellipsoid, 0.8–0.9 mm. long, with a 0.1–0.2 mm. long, narrowly clavate, apical gland, eglandular at the base. Ovary 2.5–5 × 2–3 mm., ellipsoid; style 1.5 mm. long, filiform; stigma 0.7–0.8 mm. long, papillate, narrowly ellipsoid. Capsule 3.5–5.5 × 3–4 mm., ellipsoid. Seeds 0.1–0.15 mm. in diam., cubical; testa faveolate.

Zambia. N: Luwingu, fl. & fr. 29.iii.1962, *Robinson* 5066 (K; LISC). W: Mwinilunga on Matonchi Rd., fl. & fr. 24.i.1975, *Brummitt, Chisumpa & Polhill* 14090 (K).
Also in Zaire and Angola. In boggy grassland.

14. **Sebaea teuszii** (Schinz) Taylor in Taxon **12**, 8: 294 (1963); in F.W.T.A. ed. 2, **2**: 298 (1963). —Boutique in Fl. Afr. Centr., Gentianaceae: 44 (1972). Type from Angola.
Belmontia teuszii Schinz in Viertelj. Nat. Gesellsch. Zürich. **36**: 335 (1891). —Baker & N.E. Br. in F.T.A. **4**, 1: 554 (1903). Type as above.
Tachiadenus continentalis Baker in Kew Bull. **1895**: 70 (1895). Type from Kenya.
Belmontia chionanthum Gilg in Baum, Kunene-Samb.-Exped.: 332 (1903). Type from Angola.
Exochaenum teuszii (Schinz) Schinz in Bull. Herb. Boiss., Sér. 2, **6**: 716 (1906). —De Wild., Ann. Mus. Congo. Bot., Sér. 5, **2**: 336 (1908). Type as above.
Exochaenium chionanthus (Gilg) Schinz in Bull. Herb. Boiss., Sér. 2, **6**: 716 (1906). —De Wild. in Ann. Mus. Congo, Bot., Ser. 5, **2**: 336 (1908); in Pl. Bequaert, **2**: 104 (1923). Type as above.
Belmontia teuszii var. *angustifolia* De Wild. in Bull Jard. Bot. Etat. Brux., **3**: 279 (1911); in Ann. Mus. Congo Bot., Sér. 4, **2**: 122 (1913). Type from Zaire.
Belmontia chevalieri Abbayes & Schnell in Bull. Soc. Bot. Fr. **96**: 204 (1950). Type from W. Africa.

Delicate annual herbs up to 60 cm. tall. Stem simple or branched, 4-ridged. Leaves in 4–8 pairs, 5–25 × 2–8 mm., oblong-lanceolate to linear-lanceolate, acute or subobtuse at the apex. Flowers white, pale cream to yellowish, solitary, terminal, pedicellate; pedicels 1.5–13 cm. long; calyx with 4–5 segments, 7–20 mm. long, lanceolate, acuminate, with a dorsal wing, 0.5–0.6 mm. broad near the middle. Corolla tube 10–30 × 1–2 mm., subcylindric, the lower portion enlarged in fruit, narrowed around the ovary, widening again from the point of insertion of the filaments, puberulous on the external face; lobes 5–25 × 2–12 mm., ovate-lanceolate, attenuate or apiculate at the apex. Stamens included, inserted in the middle of the corolla tube; filaments 5, subclaviform, alternate with 5 short protuberances; anthers 3.5 mm. long, narrowly ellipsoid with a triangular-rhombic to narrowly clavate apical 0.4 × 0.6 mm. long gland; basal glands 0.3 mm. long, stipitate, sometimes absent. Ovary 5.5–8 mm. long, ellipsoid; style 4–9 mm. long; stigma papillate. Capsule 7.5–10 mm. long, narrowly ovoid to ellipsoid. Seeds 0.20–0.25 mm. long, cubical; testa faveolate.

Zambia. N: Mbala Distr., Lake Chila, fl. & fr. 24.ix.1949, *Bullock* 1061 (K). W: Kitwe, fl. & fr. 20.iii.1955, *Fanshawe* 2158 (K). C: Between Serenje and Mpika, fl. & fr. 16.vii.1930, *Pole-Evans* 2910 (K; PRE).
Also in Guinea and Sierra Leone, Angola, Zaire and Tanzania. In open grass swamp, in wet places; 1200–1780 m.

15. **Sebaea grandis** (E. Mey.) Steud., Nom., ed. 2, **1**: 196 (1840); **2**: 550 (1841). —Marais & Verdoorn in

Fl. S. Afr. **26**: 189 (1963). —Taylor in F.W.T.A. ed. 2, **2**: 298 (1963). —Friedrich-Holzhammer in Merxm., Prodr. Fl. SW. Afr., **110**, Gentianaceae: 4 (1967). —Boutique in Fl. Afr. Centr., Gentianaceae: 43 (1972). —Ross in Bot. Surv. Mem. **39**: 278 (1972). TAB. **8**. Type from S. Africa.

Belmontia grandis E. Mey. in Comm. Pl. S. Afr. **2**: 183 (1837). —Baker & N.E. Br. in F.T.A. **4**, 1: 553 (1903). Type as above.

Exochaenium grande (E. Mey.) Griseb. in DC., Prodr. **9**: 55 (1845). Type as above.

Belmontia zambesiaca Baker in Kew Bull. **1894**: 25 (1894). Type: Botswana, Zambesia, in the Valley of the Leshumo, *Holub* s.n. (K).

Parasia grandis (E. Mey.) Hiern, Cat. Afr. Pl. Welw. **1**, 3: 707 (1898). Type from S. Africa.

Parasia grandis var. *major* S. Moore in Journ. Bot. **40**: 348 (1902). Type from S. Africa (Transvaal).

Exochaenium grande var. *major* (S. Moore) Schinz in Bull. Herb. Boiss., Sér. 2, **6**: 802 (1906). Type as above.

Sebaea natalensis (Schinz) Schinz in Bull. Herb. Boiss. Sér. 2, **6**: 736 (1906) non Schinz (1895). Type from S. Africa (Natal).

Exochaenium grande var. *homostylum* Hill in Kew Bull. **1908**: 338 (1908). —Hill & Prain in F.C. **4**, 1: 1904 (1909). Syntypes from S. Africa (Natal).

Exochaenium macranthum Hill in Kew Bull. **1908**: 339 (1908). Syntypes from Uganda, Angola, Malawi: *McLounie* 98 (K); 126 (K), Mozambique: *Last* s.n. (K).

Erect annual, more rarely perennial rhizomatous herbs up to 40 cm. tall. Stem unbranched or sometimes branched from the base, 4-ridged. Leaves 5–40 × 2.5–10 mm., lanceolate to ovate-lanceolate, acute or acuminate at the apex, the lower ones often reduced and scale-like. Flowers white, pale-yellow with slight-orange tinge, orange-yellow to brownish, solitary, terminal or in few flowered cymes with long branches, heterostylous: short-styled flowers with the anthers exceeding the stigma; medium-styled flowers with the anthers at the level of the stigma and long styled flowers with the stamens shorter than the stigma. Calyx segments 4–5, 7–22.5 × 1–4 (5) mm. long, lanceolate to elliptic-lanceolate, long acuminate, dorsally broadly winged, wing up to 2 mm. broad near the base narrowing towards the apex. Corolla tube 12–30 mm. long, broadly cylindrical, the lower portion enlarged in fruit, narrowed above the ovary, widening again from the point of insertion of the filaments, puberulous on the external face; lobes 5–20 × 3–14 mm., obovate-oblong to subcircular, acute and apiculate. Stamens included, inserted in the corolla tube; filaments 0.3–4.5 mm. long; anthers 0.6–1.5 mm. long, oblong with a narrowly clavate, stipitate 0.75–1.5 mm. long, apical gland and 2 small rounded basal glands. Ovary 5–6 mm. long, globose or obovoid; style 3.5–6 mm. long; stigma claviform–oblong, papillate. Capsule 7–8 mm. in diam., subglobose. Seed 0.1–0.2 mm. in diam., cubical; testa faveolate.

Botswana. SW: Near Old James Camp, fl. & fr. 5.iv.1976, *Williamson* 96 (PRE; SRGH). **Zambia**. B: Sesheke, fl. ii.1911, *Gairdner* 475 (K). N: Mbala Distr., Ndundu, fl. 23.ii.1964, *Richards* 19063 (K). C: Chakwenga, 100–129 km. E. of Lusaka, fl. & fr. 5.iii.1965, *Robinson* 6415 (K). E: Katete R., near Katete, fl. 10.iii.1957, *Wright* 475 (K). **Zimbabwe**. N: Urungwe (Hurungwe) Nat. Park, 10 km. from Makuti on Kariba Rd., fl. 18.ii.1981, *Philcox, Leppard & Dini* 8704 (K). W: Hwange Nat. Park, Ugamo Flats, SE. of Main Camp, fl. 13.iii.1969 *Rushworth* 1690 (K; SRGH). C: Rusape, St. Faiths' Mission Farm, fl. 14.iii.1958, *Norman* R23 (K). E: Nyanga Rd. to lower Inyangombi Falls, fl. 17.i.1959, *Lennon* 92 (K; SRGH). S: Bikita, fl. & fr. 9.v.1969, *Biegel* 3076 (K). **Malawi**. N: Viphya Plateau, fl. & fr. 1.iii.1948, *Benson* 1458 (BM). C: Dedza, Dzoloro, Chongoni Forest, fl. & fr. 22.iii.1967, *Salubeni* 618 (K; SRGH). S: Blantyre, fl. 16.ii.1938, *Lawrence* 658 (K). **Mozambique**. N: Niassa, Malema, Serra Merripa, fl. 4.ii.1964, *Torre & Paiva* 10433 (LISC). Z: Namagoa, Mocuba, fl. & fr. vii.1945, *Faulkner* 252 (K; PRE). T: Angonia, 11 km. of Calobue, fl. & fr. 7.iii.1964, *Torre & Paiva* 11054 (LISC). MS: Manica, base of Serra de Mavita, fl. & fr, 3.iv.1948, *Barbosa* 1551 (LISC). M: Libombos, near Namaacha, Mponduine Mt., fl. 22.ii.1955, *Exell, Mendonça & Wild* 501 (BM; LISC; SRGH).

Also in Nigeria, Cameroons and S. Africa. In open grassland, among rocks in sandy soil; 500–2100 m.

16. **Sebaea clavata** Paiva & Nogueira in Anal. Jard. Bot. Madrid **47**, 1: xxx (1990). TAB. **9**. Type: Zambia, road from Senga Hill to Mporokoso, *Robinson* 1736 (K, holotype).

Erect annual herb up to 30 cm. tall. Stem slender, sparingly branched above, 4-ridged. Leaves 15–70 × 1–2 mm., linear to linear-lanceolate, acute at the apex, the lower subrosulate. Flowers pale yellow or bright yellow in a lax 1–few flowered terminal cyme, pedicellate; pedicels up to 20 cm. long. Calyx segments 5, 8–10 mm. long, lanceolate, caudate at the apex with a dorsal wing 0.4–0.8 mm. broad below. Corolla tube 10–15 mm. long; lobes 4–5 mm. long, ovate, rounded and apiculate at the apex. Stamens included,

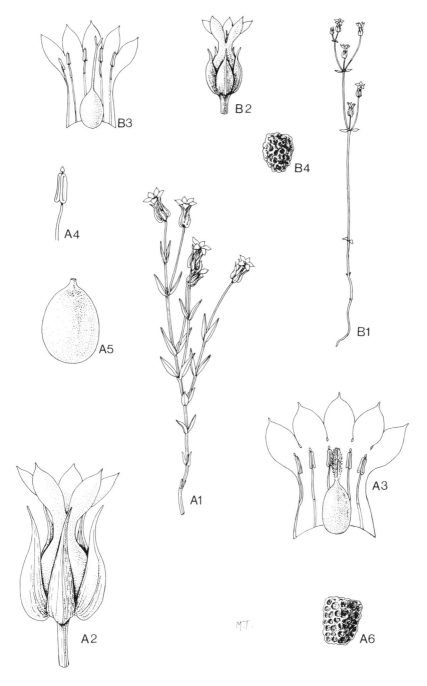

Tab. 8. A.—SEBAEA GRANDIS. A1, flowering stem (×½); A2, flower (× 3); A3, corolla opened out to show pistil and stamens (× 6); A4, stamen (× 12); A5, capsule (× 3); A6, seed (× 40), A1–6 from *Faulkner* 252. B.—SEBAEA JUNODII. B1, habit (×½); B2, flower (× 3); B3, corolla opened out to show pistil and stamens (× 3); B4, seed (× 40), B1–4 from *Richards* 9579.

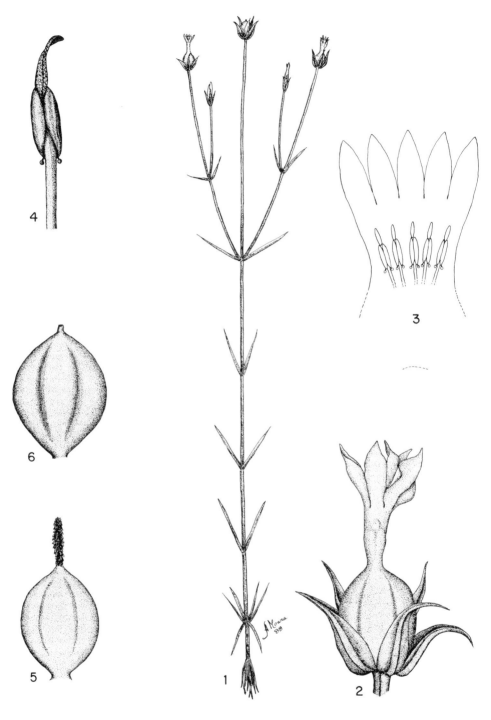

Tab. 9. SEBAEA CLAVATA. 1, habit (× $\frac{1}{3}$); 2, flower (× 3); 3, corolla opened out showing stamens (× 4); 4, stamen (× 6$\frac{1}{2}$); 5, pistil (× 3); 6, capsule (× 3), all from *Robinson* 1736.

inserted at the base of the corolla tube; filaments 1.5 mm. long; anthers ellipsoid, with a narrowly clavate, apical, 1–1.2 mm. long gland, and two minute globose, stipitate, basal glands. Ovary 4–5 × 3–4 mm., subglobose; style c. 0.4–0.5 mm. long; stigma 2–3 mm. long, clavate-papillate. Capsule 7–9 × 4–6 mm., ellipsoid; seeds 0.2–0.25 mm., cubical; testa faveolate.

Zambia. N: Mbala Distr., c. 24 km. from Senga Hill to Mporokoso, fl. & fr. 8.vi.1956, *Robinson* 1736 (K, holotype).
Not known elsewhere. In *Brachystegia* woodland, saturated marshland among taller, robust vegetation; 1600–1700 m.

17. **Sebaea pentandra** E. Mey., Comm. Pl. S. Afr.: 184 (1877). Type from S. Africa.

Var. **burchellii** (Gilg) Marais in Bothalia **7**: 464 (1961). —Marais & Verdoorn in Fl. S. Afr. **26**: 204, fig. 26, 8 (1963). Type from S. Africa (Cape Prov.).
Sebaea burchellii Gilg in Engl., Bot. Jahrb. **26**: 89 (1898). —Schinz in Bull. Herb. Boiss., Sér. 2, **6**: 724 (1906). —Hill & Prain in F.C. **4**, 1: 1076 (1909). Type from S. Africa.
Sebaea conrathii Schinz in Mitt. Geogr. Ges. Lübeck **17**: 31 (1903). —Hill & Prain in F.C. **4**, 1: 1076 (1909). Type from S. Africa (Transvaal).
Sebaea barbeyana Schinz in Viertelj. Nat. Gesellsch. Zürich **37**: 326 (1891); in tom. cit.: 32 (1903). —Gilg in tom. cit.: 101. —Baker & N.E. Br. in F.T.A. **4**, 1: 549 (1903). Type from Namibia.
Sebaea dinteri Gilg ex Dinter in Feddes Repert. **23**: 231 (1926) nom. nud.
Sebaea pentandra sensu Friedrich-Holzhammer in Merxm., Prodr. Fl. SW. Afr. **110**, Gentianaceae: 4 (1964) non E. Mey.

Erect annual herb up to 25 cm. tall. Stem slender, branched from the base, glabrous, 4-ridged. Leaves crowded near the base, the lower subrosulate, narrowed into a petiole-like base; the upper sessile, in 3–5 distant pairs; lamina 1.3–4 × 0.8–2 cm., ovate-oblong to elliptic obtuse or subacute at the apex, rounded or cuneate at the base. Flowers pale yellow to bright yellow, in terminal few–many flowered lax cymes, arranged corymb-like; pedicels 2–15 mm. long; bracts 2–4 × 1–2 mm., ovate to elliptic-lanceolate. Calyx segments 5, 3–5 mm. long, ovate-lanceolate, acute, not winged on the keel. Corolla tube 3–5 mm. long; lobes 2–3 × 0.75–1.75 mm., obovate-oblong to obovate-circular; stamens inserted 0.75 mm. below the corolla sinuses; filaments 0.5 mm. long; anthers 0.75–1.25 mm. long, apices recurved, with a round, apical, stipitate gland and 2 small stipitate basal glands. Ovary 3–3.5 × 1–2.5 mm. long, ovate; style 1.25–2.25 mm., with or without swelling above the middle; stigma 0.2–0.3 mm. clavate or sub-capitate, bilobed. Capsule 4–4.5 × 2–2.25 mm., obovoid. Seeds 0.15–0.2 mm. long, cubical; testa frilled.

Botswana. SE: Gaborone Distr., Aedume Park, fl. & fr. 6.viii.1978, *Hansen* 3423 (C; GAB; K; PRE). **Zambia**. C: Lusaka Distr., Chongwe R., N. of Kasisi (c. 30 km. NNE. of Lusaka) fl. & fr. 29.ix.1972, *Strid* 2236 (K). S: Victoria Falls, fl. & fr. 1860 *Kirk* s.n. (K). **Zimbabwe**. W: Isenzi R., fl. & fr. 1930, *Chesman* 233 (K). E: Chimedza R., c. 3.5 km. from junction with Tsungwesi R., fl. & fr. 8.x.1955, *Drummond* 4864 (K; LISC). S: Chibi, near Madzivire Dip, c. 5 km. N. of Lundi R. bridge, fl. & fr. 3.v.1962, *Drummond* 7880 (K; LISC).
Also in S. Africa, Namibia and Angola. In moist places, beside rivers, damp sandy soil between rocks in river beds; 1050 m.
S. pentandra var. *pentandra* does not differ in habitat from var. *burchellii*, but has a more southerly distribution (S. Africa: Orange Free State and Cape Province). Whereas the latter has almost leafless stems, smaller flowers (calyx segments 3–5 mm. long; corolla lobes 2–3 × 0.75–1.75 mm.), very short filaments (to 0.5 mm. long) and the style with (rarely without) a stigmatic swelling above the middle; the former usually has leafy stems, larger flowers (calyx segments 4.5–7.5 mm. long; corolla lobes 3.5–7.5 × 1.75–3.5 mm.), filaments to 2.5 mm. long and the stigmatic swelling of the style below the middle (usually very near the base).

18. **Sebaea brachyphylla** Griseb., Gen. et Sp. Gent.: 170 (1839). —Gilg in Engl., Bot. Jahrb. **26**: 100 (1898); in Mildbr., Deutsch. Zentr. Afr. Exp. 1907–1908, **2**: 544 (1913). —Hiern, Cat. Pl. Afr. Welw. **1**, 3: 705 (1898). —Taylor in F.W.T.A. ed. 2, **2**: 298, fig. 267 (1963). —Boutique in Fl. Afr. Centr., Gentianaceae: 42 (1972). Type from Madagascar.
Sebaea schimperiana Buching. ex Schweinf., Bietr. Fl. Aethiop.: 127 (1867). Type from Ethiopia.
Sebaea multinodis N.E. Br. in F.T.A. **4**, 1: 548 (1903). —Hutch. & Dalz. in F.W.T.A. **2**, 1: 182, fig. 236 (1931). Type from Cameroon.
Sebaea baumii Schinz in Mit. Geogr. Ges. Lübeck **17**: 27 (1903). Type from Angola.
Sebaea butaguensis De Wild. in Rev. Zool. Afr. 10 (Suppl. Bot.) **1**: 33 (1922); in Pl. Bequaert **2**: 106 (1923). Type from Uganda?

Erect annual herb up to 40 cm. tall. Stem slender, unbranched or branched (sometimes almost from the base), 4-ridged. Leaves, sessile, 5-nerved; lamina 5–12 × 5–14 mm., ovate-circular to subreniform, cordate at the base, obtuse of shortly apiculate at the apex. Flowers yellow or pale yellow, in terminal few to many flowered dichotomous cymes, very rarely solitary; pedicels 0.3 mm. long; bracts 2–3 mm. long, elliptic-lanceolate. Calyx segments 5, 2.5–5 × 1.5–3 mm. long, ovate-lanceolate, acute at the apex, with a short wing on the keel, broader at or above the middle. Corolla tube, 3–4 mm. long; lobes 1.5–3.5 × 0.8–1.2 mm. obovate-oblong. Stamens inserted in the throat of the corolla; filaments 0.2–0.7 mm. long; anthers 0.5–0.7 mm. long with a minute ovoid-ellipsoid apical gland, without basal glands. Ovary ellipsoid; style 3–5 mm. long, with or without a swelling above the middle; stigma subglobose more or less bilobed. Capsule 2–4 mm. long, ellipsoid. Seeds 0.25–0.3 mm. in diam., cubical; testa frilled.

Zambia. N: Kasama Distr., Mungwi, fl. & fr. 19.xi.1960, *Robinson* 4084 (K). W: Mwinilunga Distr., R. Lunga, fl. & fr. x.1937, *Milne-Redhead* 3479 (K; PRE). C: Serenje, fl. & fr. 16.x.1963, *Robinson* 5763 (K). **Zimbabwe.** C: Harare, fl. & fr. xi.1955, *Drummond* 4946 (K; LISC). E: Chimanimani, Iona Farm, fl. & fr. 15.iii.1953, *Chase* 4852. **Malawi.** N: Viphya Mt., Mzimba, fl. 12.vi.1947, *Benson* 1290 (K). S: Shire Highlands, Zambesiland, fl. & fr. 6.iv.1906, *Adamson* (K). **Mozambique.** N: Lichinga, fl. & fr. 20.i.1935, *Torre* 296 (COI; LISC). M: Maputo, Bela Vista, between Zitundo and Ponta do Ouro, fl. & fr. 3.xii.1968, *Balsinhas* 1430 (LISC).

Also in Nigeria, Cameroon, Bioko, Uganda, Kenya, Angola and Madagascar. Marshy ground, wet sand and rocky situations; 1300–1600 m.

19. **Sebaea leiostyla** Gilg in Engl., Bot. Jahrb. **26**: 97 (1898). —Schinz in Mitt. Geogr. Ges. Lübeck **17**: 32 (1903); in Bull. Herb. Boiss., Sér. 2, **6**: 727 (1906). —Baker & N.E. Br. in F.T.A. **4**, 1: 548 (1903). —Hill in Kew Bull. **1908**: 330 (1908). —Hill & Prain in F.C. **4**, 1: 1086 (1909). —Marais & Verdoorn in Fl. S. Afr. **26**: 207, fig. 28, 6 (1963). —Boutique in Fl. Afr. Centr., Gentianaceae: 40, t. 6 (1972). —Ross, in Bot. Surv. Mem. **39**: 278 (1972). Syntypes: Malawi, without precise locality *Buchanan* 270 (K; Z); *Buchanan* 200 (K).

   Sebaea polyantha Gilg in Engl., Bot. Jahrb. **26**: 95 (1898). —Schinz in Mitt. Geogr. Ges. Lübeck **17**: 47 (1903). —Hill & Prain in F.C. **4**, 1: 1087 (1909). Type from S. Africa.

   Sebaea transvaalensis Schinz in Mitt. Geogr. Ges. Lübeck **17**: 49 (1903). Type as above.

   Sebaea bequaertii De Wild. in Rev. Zool. Afr. **10** (Suppl. Bot.): 31 (1922); in Pl. Bequaert **2**: 105 (1923). Type from Zaire.

   Sebaea schimperiana auct. non Buching ex Schweinf (1867).

   Sebaea brachyphylla auct. non Griseb (1839).

Erect annual or perennial herb up to 60 cm. tall. Stem slender, unbranched or branched (sometimes from the base), 4-ridged. Leaves sessile; lamina 6–25 × 7–30 mm. ovate, ovate-circular to subreniform, membranous or leathery, cordate at the base obtuse or shortly apiculate, 5-nerved. Flowers bright yellow to orange in terminal many flowered dichotomous cymes arranged corymbosely or in paniculate corymbose, subdense or dense cymes contracted into round, almost head-like, inflorescence when young, rarely solitary flower; pedicels 1–1.8 mm. long; bracts 2–3 mm. long, ovate lanceolate-elliptic acute, recurved. Calyx segments 4–7.5 × 1.5 mm., lanceolate, to elliptic, acute or acuminate, margin hyaline, with a keel shortly winged, wing broadest below or towards the middle. Corolla tube, (3.5) 4.5–7 mm. long; lobes 5–6.5 × 2.5–3.5 mm. obovate-oblong, obtuse at the apex. Stamens inserted at the corolla-sinuses; filaments 1.75–2 mm. long; anthers 1–2.25 mm. long, oblong, with a minute apical gland and sometimes with 2 minute basal glands. Ovary ellipsoid; style 3.5–5 mm. long, with or without a swelling below the middle or near the base; stigma 1.5–2.5 mm., capitate or clavate, bilobed. Capsule ellipsoid. Seeds 0.35–0.45 × 0.25–0.3 mm., ellipsoid; testa frilled.

Zimbabwe. E: Mutare, fl. & fr. 29.iv.1953, *Chase* 4889 (BM; LISC). S: Mberengwa, fl. & fr. 3.v.1971, *Pope, Biegel & Simon* 1075 (K; LISC; PRE; SRGH). **Malawi.** N: Nyika Plateau, Kasaramba, fl. & fr. 24.v.1967, *Salubeni* 723 (K; LISC; PRE; SRGH). C: Dedza Distr., fl. & fr. 24.iv.1970, *Brummitt* 10118 (K). S: Ntcheu Distr., Lower Kirk Range, Chipusiri, fl. & fr. 17.iii.1955, *Exell, Mendonça & Wild* 974 (BM; LISC). **Mozambique.** Z: Gúruè, near Namuli Peak, fl. & fr. 9.iv.1943, *Torre* 5144 (LISC). MS: Chimanimani, fl. & fr. 17.ii.1967, *Grosvenor* 411 (K; LISC; SRGH).

Widespread throughout eastern, south and tropical Africa. In grassland, in damp soil over granite and rocky outcrops; 1300–2140 (3500 m.).

20. **Sebaea longicaulis** Schinz in Bull. Herb. Boiss. **2**: 219 (1894); in Mitt. Geogr. Ges. Lübeck **17**: 42 (1903); in Bull. Herb. Boiss., Sér. 2, **6**: 727 (1906). —Gilg in Engl., Bot. Jahrb. **26**: 94 (1898). —Hill & Prain in F.C. **4**, 1: 1083 (1909). —Taylor in F.W.T.A. ed. 2, **2**: 298 (1963). —Marais & Verdoorn

in Fl. S. Afr. **26**: 204, fig. 28, 2 (1963). —Ross in Bot. Surv. Mem. **39**: 278 (1972). Type from S. Africa.

   Sebaea crassulifolia var. *lanceolata* Schinz in Viestelj. Nat. Ges. Zürich **37**: 323 (1891). Type from S. Africa (Natal).

   Sebaea macrosepala Gilg in Engl., Bot. Jahrb. **26**: 91 (1898). Type from S. Africa (Transvaal).

   Sebaea oreophila Gilg in Engl., Bot. Jahrb. **30**: 377 (1901). —Baker & N.E. Br. in F.T.A. **4**, 1: 547 (1903). Type: Malawi, *Göetze* 945 (K, isotype).

   Sebaea macowanii Gilg ex Schinz in Mitt. Geogr. Ges. Lübeck **17**: 47 (1903). —Hill & Prain, tom. cit.: 1087 (1909). Type from S. Africa.

   Sebaea grandiflora Schinz in Mitt. Geogr. Ges. Lübeck **17**: 44 (1903); in Bull. Herb. Boiss., Sér. 2, **6**: 726 (1906). —Hill & Prain tom. cit.: 1804 (1909). Type from S. Africa (Transvaal).

   Sebaea erecta Hill in Kew Bull. **1908**: 328 (1908). —Hill & Prain tom. cit.: 1083 (1090). Type from S. Africa (Transvaal).

Erect herb up to 60 cm. tall. Stem unbranched or branched, 4-ridged. Leaves sessile; lamina 18 × 15 mm., elliptic-lanceolate to subcircular-cordate or subcircular-reniform, rounded or acute, or sometimes mucronate at the apex, sometimes coriaceous, the upper smaller and more distant, 5-nerved. Flowers yellow, in few flowered lax or compact cymes, terminal, solitary or corymbose with very few bracts; bracts ovate-lanceolate to linear or linear-subulate, erect. Calyx segments 5, 7–11 × 2–5.5 mm. long, lanceolate to elliptic, acute, margin hyaline, keeled or with a dorsal wing broadest at the middle or near the base. Corolla tube 5.5–9 mm. long; lobes 5.5–11 × 2.5–5.5 mm., broadly elliptic to oblong-obovate or obovate-circular, rounded, clawed. Stamens inserted at the corolla sinuses; filaments 1–4 mm. long; anthers narrowly ellipsoid, 2–4.25 mm. long, with a minute apical gland and sometimes with 2 minute basal glands. Ovary ellipsoid, 3.5–4 × 1.5–2 mm.; style 6.5–9.5 mm. long, without a stigmatic swelling and then the stigma small, capitate bilobed, or with a swelling near the base and then a large clavate stigma with 2 broad tongue-shaped lobes. Capsule 5–6 × 2.5–3 mm., ellipsoid. Seeds 0.35–0.4 × 0.3 mm., ellipsoid or cubical; testa frilled.

Zimbabwe. E: Stapleford, fl. & fr. 22.ii.1964, *Wild* 14512 (K). C: Marondera, fl. & fr. 9.xi.1950, *Corby* 706 (K). **Malawi**. N: Nyika Plateau, fl. & fr. 14.viii.1946, *Brass* 17226 (BM; K). S: Mulanje Distr., Lichenya Plateau, fl. & fr. 26.vi.1946, *Brass* 16434 (BM; K). **Mozambique**. MS: Tsetserra, fl. & fr. 21.vi.1972, *Biegel* 3965 (K; LISC; SRGH).

Also in the Cameroons and S. Africa. In grasslands; 1700–2300 m.

21. **Sebaea sedoides** Gilg in Engl., Bot. Jahrb. **26**: 98 (1898). Type from S. Africa.

Var. **confertiflora** (Schinz) Marais in Bothalia **7**: 464 (1961). —Marais & Verdoorn in Fl. S. Afr. **26**: 210 (1963). —Ross in Bot. Surv. Mem. **39**: 278 (1972). Type from S. Africa (Transvaal).

   Sebaea confertiflora Schinz in Mitt. Geogr. Ges. Lübeck **17**: 51 (1903). —Hill & Prain in F.C. **4**, 1: 1082 (1909). Type as above.

Perennial herbs up to 65 cm. tall. Stem annual from a rootstock, erect or ascending unbranched or branched, 4-ridged. Leaves 18 × 20 mm. reniform-circular or broadly cordate, rounded to subobtuse, sometimes apiculate at the apex. Flowers yellow, butter yellow or white, in a very dense corymb sometimes much contracted and head-like with relatively large bracts; bracts ovate to lanceolate, or narrowly obovate to oblanceolate, obtuse to acute. Calyx segments 5, 3–8 × 1.75–2.75 mm., lanceolate to elliptic-lanceolate or linear-oblong. Corolla tube (3.5)4.5 – 7.5 mm. long; lobes 4.5–7.25 × 2.75 mm. lanceolate. Stamens inserted at the corolla sinuses; filaments up to 1.25 mm. long; anthers 1–2.5 mm., narrowly ellipsoid, with a minute apical gland and sometimes with 2 minute basal glands. Ovary 2.5–3.5 × 0.75, ellipsoid; style 4–7.25 mm. long, with a small or medium swelling below the middle, rarely without a swelling; stigma capitate or capitate-clavate. Capsule 3.5–4 × 1.5–2 mm. ellipsoid. Seeds 0.25–0.3 mm. ellipsoid, numerous, cubical; testa frilled.

Zimbabwe. E: Nyanga (Inyanga), fl. & fr. 30.iii.1949, *Chase* 1632 (BM; K; PRE). **Mozambique**. MS: Barue Mt., c. 15 km. Vila de Gouveia, fl. & fr. 28.iii.1966, *Torre & Correia* 15487 (LISC).

Also in Swaziland and S. Africa. Marshy ground and grasslands; 1300–1500 m.

S. sedoides var. *sedoides* (Swaziland and S. Africa) is generally more robust and more branched than var. *confertiflora* and var. *schoenlandii* (Swaziland and S. Africa) and the inflorescences are paniculate-corymbose, whereas the others have the flowers arranged in densely corymbose head-like inflorescences. *S. sedoides* var. *confertiflora* is distinguishable from var. *scheonlandii* by having the filaments up to 1.25 mm. long and inserted in the corolla-sinuses, whereas the latter has the filaments up to 0.5 mm. long and inserted below the corolla-sinuses.

5. SWERTIA L.

Swertia L., Sp. Pl.: 226 (1753).

Annual or perennial herbs. Stems erect, ascending or straggling; simple or branched, 4-ridged. Cauline leaves opposite in distant pairs, sessile or very shortly petiolate; basal leaves (sometimes soon deciduous) narrowed into a petiole. Flowers 4–5 merous in lax, few–many flowered cymes, rarely solitary. Calyx 4–5 lobed with a very short tube, lobes linear, lanceolate, ovate to obovate. Corolla 4–5 lobed, white, blue or yellow, with a very short tube; lobes erect, with 1–2 basal glandular nectaries within. Stamens 4–5, inserted in the sinuses of the corolla, shorter than the lobes; anthers small. Ovary ellipsoid, unilocular; ovules numerous; styles short or obsolete; stigma bilobed. Capsule oblong or lanceolate in outline, splitting septicidally, bivalved. Seeds numerous subglobose, compressed, faveolate or frilled.

A genus of about 100 species occurring throughout Europe, Africa and Asia.

1. Flowers 4-merous - - - - - - - - - - - - - - - 2
– Flowers 5-merous - - - - - - - - - - - - - - - 4
2. Leaves widening upwards; calyx lobes subspathulate, 0.9–1.2 mm. broad, subequal, obtuse or
    subobtuse at the apex - - - - - - - - - - - 12. *intermixta*
– Leaves narrowing upwards; calyx lobes elliptic to linear, 0.25–0.5 mm. broad, very unequal, acute
    at the apex - - - - - - - - - - - - - - 3
3. Leaves becoming longer upwards; calyx lobes as long as or longer than the corolla, linear, up to 7
    mm. long - - - - - - - - - - - 11. *tetrandra*
– Leaves shortening upwards; calyx lobes much shorter than the corolla, elliptic to elliptic-
    lanceolate, up to 3.5 mm. long - - - - - - - - - 9. *eminii*
4. Corolla lobes with one nectary - - - - - - - - - 5
– Corolla lobes with two nectaries - - - - - - - - 6
5. Leaf lamina ovate-lanceolate to triangular-ovate, 0.7–2.8 cm. broad; capsule 6.5–7 × 2.5–5
    mm. - - - - - - - - - - - - - 2. *kilimandscharica*
– Leaf lamina lanceolate to linear-lanceolate, 0.5–0.8 cm. broad; capsule 12–15 × 5–7
    mm. - - - - - - - - - - - 1. *johnsonii*
6. Cauline leaves sessile; stems unbranched or branching at the apex - - - 7
– Cauline leaves narrowed at the base into a short petiole; stems usually branched from the
    base - - - - - - - - - - - - - - 11
7. Perennial, ascending or erect herbs; upper cauline leaves reflexed
    or erect-patent - - - - - - - - - - - - 8
– Annual, erect herbs; upper cauline leaves erect - - - - - - 9
8. Basal leaves longer than the cauline ones - - - - 4. *quartiniana*
– Basal leaves shorter than the cauline ones - - - - 3. *fwambensis*
9. Upper leaves folded lengthwise, spreading - - - - 7. *curtioides*
– Upper leaves not folded, ascending or only those of inflorescence
    base spreading - - - - - - - - - - - 10
10. Leaf lamina linear, 1–2 mm. broad; flowers in terminal and axillary cymes arranged in a very
    narrow panicle; calyx lobes 6–9 mm. long; anthers pale yellow - - 6. *pleurogynoides*
– Leaf lamina linear-lanceolate, 1.5–7 mm. broad; flowers in terminal cymes arranged in a
    subcorymbose inflorescence; calyx lobes 2–7 mm. long;
    anthers brown - - - - - - - - - 5. *welwitschii*
11. Flowers very small, 3–5 (7) mm. long; capsule, 4–5 × 2.5–3 mm. - - - 9. *eminii*
– Flowers larger, 5–15 mm. long; capsule 7–13 × 3–4 mm. - - - 12
12. Cauline leaves fleshy or rather thick; corolla white or more or less deeply tinged with purple;
    capsule 8–13 × 3.5–4 mm. - - - - - - - - - 10. *abyssinica*
– Cauline leaves not fleshy, thin; corolla pale blue (whitish in dried material); capsule 7–10 ×
    3–3.5 mm. - - - - - - - - - - 8. *usambarensis*

1. **Swertia johnsonii** N.E. Br. in F.T.A. **4**, 1: 572 (1903). Type: Malawi, Lake Malawi, *Johnson* s.n. (K).

Perennial erect herbs up to 100 cm. tall. Stem 3–5 mm. in diam., branched, densely leafy to the base more or less 4-ridged, dark purple. Leaves sessile in several distant pairs; lamina 2–8.5 (10) × 0.5–0.8 cm., lanceolate to linear-lanceolate, acute at the apex, truncate at the base. Flowers 5-merous, mauve, lavender, purple, bluish-purple, blue to pale violet in terminal and axillary few to many-flowered cymes arranged in a narrow panicle; pedicels 6–12 mm. long. Calyx divided almost to the base; lobes 5–6 mm. long, lanceolate or ovate-lanceolate tapering to an acute point. Corolla with a very short tube; lobes 10–15 × 5.5–8 mm., elliptic-lanceolate, acute at the apex with a large fringed nectary within, at the base. Stamens inserted in the sinuses; filaments 5–6 mm. long; anthers 1–1.2 mm.

narrowly-ellipsoid, shortly apiculate. Ovary 3–3.5 × 1.5–2 mm., narrowly-ellipsoid, acutely tapering to the stigma, but scarcely with a distinct style; stigma bilobed, subsessile, small, ovate, subobtuse. Capsule 12–15 × 5–7 mm., narrowly-ellipsoid, attenuate-rostrate at the apex. Seeds 0.5–0.6 × 0.4 mm. subglobose, to ellipsoid, frilled.

Malawi. N: Nyika Plateau, fl. & fr. 13.vii.1946, *Brass* 17149 (BM; K; PRE).
Not known elsewhere. Damp submontane grassland; 2300–2700 m.

2. **Swertia kilimandscharica** Engl. in Phys. Abhand. Kon. Akad. Wiss. Berlin 1891, **2**: 339 (1892). —Baker & N.E. Br. in F.T.A. **4**, 1: 573 (1903). —Boutique in Fl. Afr. Centr., Gentianaceae: 12, fig. 1F (1972). Type from Tanzania.
Swertia erosula N.E. Br. in F.T.A. **4**, 1: 572 (1903). Type as above.
Swertia schimperi sensu Oliv. in Trans. Linn. Soc. Ser. 2, **2**: 342 (1878) non Griseb. (1845).

Perennial erect herb up to 1 m. tall. Stem 2–3 mm. in diam., branched, densely leafy to the base, slightly 4-ridged. Leaves sessile; lamina 2–12 × 0.7–2.8 cm., broadly ovate-lanceolate to triangular-ovate, rounded to cordate or subamplexicaul at the base, gradually tapering from near the base to a very acute point. Flowers 5-merous, pale lavender to whitish-purple pink, in terminal and axillary few to many-flowered cymes, arranged in a narrow panicle; pedicels slender 0.8–3.5 cm. long. Calyx divided almost to the base; lobes, 5.5–10 × 2–2.25 mm., lanceolate, attenuate to a fine point at the apex. Corolla with a very short tube; lobes 8.5–13 × 4–6 mm., elliptic, obtuse at the apex, with a ciliate nectary within, at the base. Stamens inserted in the sinuses; filaments 3–4 mm. long; anthers 1–1.5 mm. long, narrowly-ellipsoid, shortly apiculate. Ovary 3–4 × 1.5–2 mm., ovate lanceolate in outline, tapering to the stigma, but scarcely with a distinct style; stigma bilobed subsessile, small. Capsule 6.5–7 × 2.5–5 mm., oblong in outline, attenuate rostrate at the apex. Seeds 0.5–0.6 mm. in diam., globose, frilled.

Malawi. N: Rumphi Distr., Nyika Plateau, Lake Kaulime, fl. & fr. 16.v.1970, *Brummitt* 10808 (K).
Also in Rwanda, Burundi, Uganda, Kenya and Tanzania. Edge of lakes in swampy ground; 2100–3000 m.

3. **Swertia fwambensis** N.E. Br. in F.T.A. **4**, 1: 574 (1903). Type from Tanzania.

Perennial erect herb up to 30–50 cm. tall. Stem solitary, unbranched or branched at the apex, sometimes several stems branched at the apex. Leaves in 4–6 distant pairs, sessile, the upper, usually longitudinally folded, spreading or deflexed; the basal ones smaller passing into scales at the base, erect or ascending; lamina 20–45 (50) × 5–11 mm., lanceolate to linear-lanceolate, subacute at the apex. Flowers 5-merous, pale, or dark blue, in terminal and axillary many-flowered cymes, arranged in a flat-topped corymbose inflorescence; pedicels slender, 1.2–4 mm. long. Calyx divided almost to the base, lobes 4–6.5 × 1–1.5 mm., oblong, unequal, subacute. Corolla with a very short tube; lobes 8–12 × 2–3 mm., oblong-elliptic, obtuse or rounded at the apex, with a pair of fimbriate oblong nectaries within, near the base. Stamens inserted in the sinuses; filaments 5–5.5 m. long, filiform; anthers 1–1.5 mm. long, narrowly-ellipsoid, dark brown. Ovary 4–5 × 1.5 mm., oblong in outline narrowed into a short stout style; stigma sessile, lobes broader than long, papillate. Capsule 6–6.5 × 2–2.5 mm., oblong in outline. Seeds 0.3–0.5 mm. in diam., subglobose; testa faveolate.

Zambia. N: Mbala Distr., Kali Dambo, fl. & fr. 26.i.1952, *Richards* 541 (K).
Also in Tanzania. In wet peaty soil among grass; 1525–1740 m.

4. **Swertia quartiniana** A. Rich., Tent. Fl. Abyss. **2**: 56 (1850). —Baker & N.E. Br. in F.T.A. **4**, 1: 574 (1903). —Taylor in F.W.T.A., ed. 2, **2**: 299 (1963). —Boutique in Fl. Afr. Centr., Gentianaceae: 11 (1972). Type from Ethiopia.

Rhizomatous ascending herb up to 150 cm. tall. Stem stout, 2–6 mm. diam., unbranched or sparsely branched at the apex. Leaves in distant pairs, 20–85 × 5–20 mm., the upper sessile, the basal longer than the cauline ones, oblong to ovate-oblong, narrowed into a short petiole; the upper patent to subpatent, ovate to elliptic-lanceolate, acute or subobtuse at the apex. Flowers 5-merous, white, tipped with blue or with purple veins, pale blue, or with fine blue mauve or dark blue lines, in terminal and axillary many-flowered cymes, arranged in subcorymbiform cymes; pedicels very variable; bracts 4–15 mm. long, linear-lanceolate. Calyx divided almost to the base; lobes unequal, 2–5 mm. long, oblong to oblong-obovate. Corolla with a very short tube; lobes 8–13 × 3–3.5 mm., obovate-oblong, obtuse or rounded at the apex with a pair of fimbriate nectaries within, near the base. Stamens inserted in the sinuses; filaments 4.5–6 mm. long; anthers 1–1.5 mm. long, narrowly-ellipsoid, dark brown. Ovary 4–5 × 2–2.5 mm., narrowly-ellipsoid, narrowed into a short stout style; stigma bilobed, papillate. Capsule 7–9 × 3–3.5 mm., narrowly ellipsoid. Seeds, 0.35–0.4 mm. in diam., subglobose; testa faveolate.

Zambia. W: c. 29 km. E. of Mwinilunga along road to Solwezi, fl. & fr. 22.xi.1972, *Strid* 2617 (K).
Zimbabwe. E: Nusa Mt., fl. & fr. iii.1935, *Gilliland* 1601 (BM).
Widespread throughout Cameroon, Zaire, Ethiopia, Kenya and Tanzania. Found in wet ground among tall grass, in marsh or banks of small streams.

5. **Swertia welwitschii** Engl., Hochgebirgsfl. Trop. Afr.: 339 (1892). —Gilg in Engl., Bot. Jahrb. **26**: 110 (1898). —Marais & Verdoorn in Fl. S. Afr. **26**: 240 (1963). —Boutique in Fl. Afr. Centr., Gentianaceae: 10, fig. 1B (1972). —Ross in Bot. Surv. Mem. **39**: 279 (1972). Type from Angola.
   Adenopogon stellarioides Welw., Syn. Expl.: 27 (1862) nomen nudum.
   Swertia stellarioides Ficalho [Pl. Uteis: 225 (1884) nomen nudum] ex Hiern, Cat. Welw. Afr. Pl. **1**, 3: 711 (1898). —Baker & N.E. Br. in F.T.A. **4**, 1: 581 —Hill & Prain in F.C. **4**, 1: 1119 (1909). Type as above.
   Swertia sharpei N.E. Br. in F.T.A. **4**, 1: 581 (1904). Type: Malawi, Zomba Mt., *Sharpe* 157 (K, holotype).

Annual, erect herb up to (12) 25–80 cm. tall. Stem slender, 1–3 mm. in diam., unbranched or branching at the apex, rarely from the base. Basal leaves usually soon disappearing; cauline leaves erect or suberect; leaf lamina 15–45 × 1.5–7 mm., linear-lanceolate, subobtuse or subacute at the apex; basal and sub-basal shorter and broader than the cauline ones, usually narrowed at the base into a short petiole-like claw, all 3-nerved from the base. Flowers 5-merous, pure white, cream or white or with dark purple veins, terminal in the axil of the upper leaves or in 1–many flowered cymes, arranged in a subcorymbose inflorescence; bracts 1–10 mm. long, linear; pedicels 1.5–4 cm. long. Calyx divided almost to the base; lobes slightly unequal, 2–7 × 1.5 mm., linear-lanceolate or linear-oblong, acute or subacute at the apex. Corolla divided almost to the base; tube 0.75 mm. long; lobes 6–10 × 2.5–4 mm., oblong-obovate to elliptic, rounded at the apex, each lobe with a pair of fimbriate nectaries within, near the base. Stamens inserted in the sinuses; filaments 3.5–5 mm. long; anthers 0.5–1.5 × 0.25–0.5 mm., narrowly-ellipsoid, brown. Ovary 4–6 × 2.5 mm., narrowly-ellipsoid, slightly compressed, narrowed into a short stout style; stigma bilobed, papillate. Capsule (5) 6–10 × 1.5–3 mm., narrowly-ellipsoid, slightly exserted, splitting into two valves. Seeds subglobose, numerous, 0.3 mm. in diam.; testa faveolate.

Zambia. N: Kasama Distr., Mungwi, fl. & fr. 24.vi.1960, *Robinson* 3766 (K). W: Kamwedzi, fl. & fr. 21.vi.1953, *Fanshawe* 115 (K). E: Lunkwakwa, fl. & fr. 31.ix.1966, *Mutimushi* 1447 (K). **Zimbabwe**. N: Makonde (Lomagundi), fl. & fr. May, *Archdale* 5173 (K). C: Rusape, fl. & fr. 5.ii.1949, *Munch* in GHS.22655 (K; LISC). E: Nyanga Distr., Inyangani Farm., fl. & fr. 30.iii.1949, *Chase* 1630 (BM; K; LISC). **Malawi**. N: Mzuzu, Lusangadzi Forest, fl. & fr. 26.iv.1967, *Salubeni* 673 (K; LISC; SRGH). C: Dedza, Chongoni Forestry School, fl. & fr. 10.v.1957, *Salubeni* 700 (K; LISC; PRE; SRGH). S: Zomba, fl. & fr. 15.vii.1955, *Jackson* 1733 (K). **Mozambique**. N: Niassa Province, Macanhelas, fl. & fr. 3.vii.1970, *Phillips* 85 (K; SRGH). Z: road from Molumbo to Milange c. 2 km. from Molumbo, fl. & fr. 4.vii.1970 *Bowbrick* B5 (b) (LISC; SRGH).
Widespread throughout Uganda, Kenya, Tanzania, Angola and S. Africa. Wet meadow, growing among grass in marsh along river banks; 1350–1850 m.

6. **Swertia pleurogynoides** Baker in Kew Bull. **1898**: 158 (1898). —Baker & N.E. Br. in F.T.A. **4**, 1: 582 (1904). TAB. **10**. Syntypes: Malawi; between Kondowe and Karonga, 2000–6000 ft. *Whyte* s.n. (K); Shire Highlands, *Buchanan* s.n. (K).

Annual erect herb up to more or less 40 cm. tall. Stem slender, branched at the apex; 4-ridged. Leaves in distant pairs, erect or suberect; lamina 2.5–3.5 × 1–2 mm., linear. Flowers 5-merous, white, in terminal and axillary few–many-flowered cymes, arranged in a very narrow panicle. Calyx divided almost to the base; lobes 6–9 mm. long, linear, acute at the apex. Corolla with a very short tube; lobes 7–8 × 2–2.5 mm., oblong to oblong-lanceolate, acute at the apex, with a pair of fimbriate nectaries within, near the base. Stamens inserted in the sinuses; filaments 3 mm. long, filiform; anthers 0.5–0.55 mm., long, subglobose to narrowly ellipsoid, pale yellow. Ovary c. 5 × 1 mm. ellipsoid, compressed, narrowed into a very short stout style; stigma subsessile. Capsule 7–9 × 2.5–3 mm., ellipsoid, attenuate-rostrate at the apex. Seeds 0.2–0.3 mm. in diam., subglobose; testa faveolate.

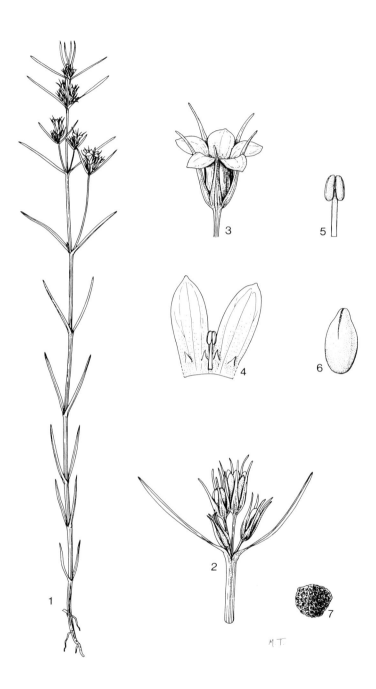

Tab. 10. SWERTIA PLEUROGYNOIDES. 1, habit (× ½); 2, part of inflorescence (× 2); 3, flower (× 2); 4, part of corolla opened out (× 3); 5, stamen (× 6); 6, capsule (× 3); 7, seed (× 40), all from *Exell, Mendonça & Wild* 932.

Malawi. S: Ntcheu Distr., Lower Kirk Range, Chipusiri, fl. & fr. 17.iii.1955, *Exell, Mendonça & Wild* 932 (BM; LISC). **Mozambique**. N: Niassa Province, Lichinga Posto Zootecnico, fl. & fr. 28.ii.1964, *Torre & Paiva* 10964 (LISC).

Known only from Malawi and Mozambique. In wet grasslands.

7. **Swertia curtioides** Gilg in Engl., Bot. Jahrb. **30**: 379 (1902). —Baker & N.E. Br. in F.T.A. **4**, 1: 580 (1904). Type from Tanzania.

Annual erect herb up to 30 cm. tall. Stem slender unbranched or branched from the base, 4-ridged. Leaves 4–30 × 2–7 mm., decreasing towards the base; the upper sessile, ovate-lanceolate, acute ot subacute at the apex, folded lengthwise, spreading; the basal ones smaller, ovate to ovate-oblong, obtuse at the apex, subpetiolate. Flowers 5-merous, whitish, cream-white or with purple veins, in terminal corymbiform cymes, pedicellate; pedicels very variable. Calyx divided to the base; lobes 1.5–4 × 1–1.2 mm., oblong-elliptic, obtuse at the apex. Corolla with a short tube; lobes 7–10 × 3–4 mm., elliptic to obovate, acute or obtuse at the apex, with a pair of fimbriate nectaries within, near the base. Stamens inserted at the sinuses; filaments 4–4.5 mm. long; anthers 1–1.5 mm., narrowly ellipsoid, dark-brown. Ovary 4–5 × 2 mm. oblong-ellipsoid, narrowed into short stout style; stigmas bilobed, papillate. Capsule 10–12 × 3.5–4 mm., narrowly ellipsoid. Seeds 0.3–0.4 mm. in diam. subglobose; testa faveolate.

Malawi. N: Nyika Plateau, fl. & fr. 14.vii.1946, *Brass* 17223 (BM; K). C: Dedza Mt., fl. & fr. 20.iii.1955, *Exell, Mendonça & Wild* 1089 (BM; LISC). S: Mulanje Mt., fl. & fr. 6.vi.1970, *Brummitt* 9629 (K). **Mozambique**. Z: Gúrùe Mt., fl. & fr. 9.iv.1943, *Torre* 5129 (LISC).

Also in Tanzania. On damp flat rocks amongst grass; 1980–2300 m.

8. **Swertia usambarensis** Engl., Pflanzenw. Ost-Afr. **C**: 314 (1895). —Baker & N.E. Br. in F.T.A. **4**, 1: 580 (1904). —Boutique in Fl. Afr. Centr., Gentianaceae: 9, fig. 1D (1972). Type from Tanzania.
    Swertia polyantha Gilg in Engl., Bot. Jahrb. **26**: 109 (1899). Type from Ethiopia.
    Swertia whytei N.E. Br. in F.T.A. **4**, 1: 579 (1904). Type from Uganda.

Annual erect herb up to 50 cm. tall. Stem slender, sometimes woody at the base, usually branched from the base, rarely unbranched or branched at the apex only, 4-ridged. Leaves thin, not fleshy; lamina 10–30 (40) × 2–7 (20) mm.; cauline ones lanceolate, attenuate at the base into a short petiole, acute or subacute at the apex, patent; basal ones, ovate-elliptic to broadly elliptic, the uppermost shorter and narrowing passing to bracts. Flowers 5-merous, pale blue (whitish in dried material) in terminal few- or many-flowered corymbiform cymes, pedicellate; pedicels very variable. Calyx divided to the base; segments 1–4 × 0.7–2 mm., unequal elliptic. Corolla with a very short tube; lobes 5–12 × 3.5–6 mm., obovate-elliptic, obtuse at the apex, with a pair of fimbriate nectaries within, near the base. Stamens inserted at the sinuses; filaments 4–6 mm. long; anthers 1–1.3 mm. long, narrowly ellipsoid. Ovary 4–5 × 2–2.5 mm., narrowly ellipsoid, narrowed into a short stout style; stigma bilobed, papillate. Capsule 7–10 × 3–3.5 mm., narrowly-ellipsoid. Seeds 0.3–0.4 mm. long; testa faveolate.

Zimbabwe. E: Nyanga Distr., Troutbeck, North Downs, fl. & fr. 8.iii.1981, *Philcox Leppard, Dini & Urayai* 8930 (K). **Malawi**. N: Nyika Plateau, fl. & fr. 13.v.1970, *Brummitt* 10654 (K). C: Dedza Distr., Chongoni Forest Reserve, fl. & fr. 25.iv.1970, *Brummitt* 10151 (K). **Mozambique**. N: Maniamba, Jeci mountain near Malulo, about 60 km. of Lichinga, fl. & fr. 3.iii.1964, *Torre & Paiva* 10996 (LISC).

Widespread throughout Uganda, Zaire, Kenya, Rwanda and Ethiopia. Grasslands; 1800–2500 m.

9. **Swertia eminii** Engl., Pflanzenw. Ost-Afr. **C**: 314 (1895). —Baker & N.E. Br. in F.T.A. **4**, 1: 580 (1904). —Boutique in Fl. Afr. Centr., Gentianaceae: 3, fig. 1C (1972). Type from Tanzania.
    Swertia tshitirungensis De Wild. in Rev. Zool. Afr., **10** (Suppl. Bot.): 4 (1922); in Pl. Bequaert **2**: 115 (1923). Type from Zaire.

Annual, erect herb. up to 50 cm. tall. Stem slender, erect sometimes ligneous at the base, branching from the base, more rarely branching at the apex. Leaves in distant pairs; the upper 10–30 × 2.5–5 mm., oblong or lanceolate, obtuse at the apex; narrowed into a petiole-like base; the lower ovate to elliptic, shortly petiolate. Flowers 4–5 merous, whitish or cream-white, in terminal and axillary 1–5 flowered corymbiform cymes, pedicellate; pedicels very variable. Calyx divided almost to the base; lobes very unequal, 1–3.5 × 0.25–0.5 mm., elliptic to elliptic-lanceolate acute at the apex. Corolla divided almost to the base; tube very short; lobes 3–5 × 2–3 mm., elliptic, obtuse at the apex, with a pair of fimbriate nectaries within, near the base. Stamens inserted in the sinuses; filaments 3–3.5 mm. long; anthers 0.7–0.8 mm. long, oblong-ellipsoid. Ovary 3.5–4.5 mm., ellipsoid or oblong, narrowed into a short stout style; stigma subsessile, bilobed. Capsule 4–5 × 2.5–3 mm., ellipsoid to narrowly so, attenuate rostrate at the apex. Seeds 0.3–0.4 mm. in diam., subglobose; testa faveolate.

Zambia. N: Mbala to Uningi Pan, fl. & fr. 17.iv.1963, *Richards* 8101 (K). **Malawi**. N: Nthalire Rd., descending from Nyika, fl. & fr. 26.iv.1973, *Pawek* 6612 (K). **Mozambique**. MS: Tsetserra Mts., fl. & fr. 3.iii.1954, *Wild* 4474 (K; LISC).
Widespread throughout Cameroon, Sudan, Zaire, Rwanda, Burundi, Uganda, Kenya and Tanzania. Edge of woodland in very long grass; 1450–2450 m.

10. **Swertia abyssinica** Hochst. in Flora **1844**: 28 (1844). —Baker & N.E. Br. in F.T.A. **4**, 1: 578 (1904).
—Taylor in F.W.T.A. ed. 2, **2**: 299 (1963). Type from Ethiopia.
    Swertia clarenceana Hook. f. in Journ. Linn. Soc. **6**: 16 (1862); op. cit. **7**: 207 (1864). —Baker & N.E. Br. in F.T.A. **4**, 1: 578 (1904). —Hutch. & Dalz., F.W.T.A. **2**, 1: 183 (1931). Type from Bioko.
    Swertia subalpina N.E. Br. in F.T.A. **4**, 1: 578 (1904). —Hutch. & Dalz., F.W.T.A. **2**, 1: 183 (1931). Type from Cameroon.
    Swertia dissimilis N.E. Br. in F.T.A. **4**, 1: 579 (1904). —Hutch. & Dalz., F.W.T.A. **2**, 1: 183 (1931). Type as above.
    Swertia porphyrantha Baker in Kew Bull. **1898**: 159 (1898). Type: Malawi, Nyika Plateau, *Whyte* 147 (K, holotype).
    Swertia lastii Engl., Hochgbirgsfl. Trop. Afr.: 337 (1892). —Baker & N.E. Br. in F.T.A. **4**, 1: 579 (1904). Type as above.

Annual, erect herb, branched from the base or middle up to 65 cm. tall. Stem weak, slender, slightly, 4-ridged. Leaves in several distant pairs, not crowded at the base, fleshy or rather thick, narrowed at the base into a short petiole; lamina 8.5–25 × 3.6–10.5 mm., obtuse or subacute at the apex, spreading. Flowers 5-merous, white or cream or white with blue lines, in terminal or axillary 2–5-flowered cymes, arranged corymbosely, rarely solitary cyme. Calyx divided almost to the base; lobes 3–6 mm. long, oblanceolate to ovate-spathulate. Corolla with a very short tube; lobes 8–13 × 4–6 mm., elliptic-oblong or oblong-ovate, obtuse at the apex, with a pair of fringed nectaries within, near the base. Stamens inserted in the sinuses; filaments 0.5–0.6 mm. long, filiform; anthers 0.45–0.50 × 0.25 mm., narrowly ellipsoid. Ovary 5–6 × 2 mm., ellipsoid, subobtuse; stigma sessile bilobed, lobes broad, obtusely rounded. Capsule 8–13 × 3.5–4 mm., ellipsoid, attenuate-rostrate at the apex. Seeds 0.25–0.3 mm. in diam., globose; testa faveolate.

Malawi. N: Chitipa Distr., Nyika Plateau, fl. & fr. 10.iv.1969, *Pawek* 2067 (K). S: Mulanje Mt., fl. & fr. 19.iv.1957, *Chapman* 451 (BM; PRE).
Also in Ethiopia, Cameroon and Bioko. In grassland, peat soil; 1500–3048 m.

11. **Swertia tetrandra** Hochst. in Flora: 28 (1844). —A. Rich., Tent. Fl. Abyss. **2**: 57 (1850). —Baker & N.E. Br. in F.T.A. **4**, 1: 582 (1904). Type from Ethiopia.

Annual, erect herb up to 18 cm. long tall. Stem slender, unbranched or branching at the apex, 4-ridged. Leaves in distant pairs, narrowing upwards; the basal 6–11 × 3.5–5 mm., obovate to elliptic-oblong, obtuse at the apex, narrowed into a short petiole, longest on upper stem, 9–13 × 0.3–0.5 mm., linear-lanceolate, sessile, acute at the apex. Flowers 4-merous, whitish, in terminal corymbiform cymes, pedicellate; pedicels very variable. Calyx divided almost to the base; lobes very unequal, as long or longer than the corolla, up to 7 × 0.5 mm. long, elliptic to linear, acute and arched at the apex. Corolla with a short tube; lobes 6–7 × 1.5–2 mm., linear-lanceolate, acute or subacute at the apex, with a pair of fimbriate nectaries within, near the base. Stamens inserted at the sinuses; filaments 2–3 mm. long; anthers 0.4 mm. long, narrowly ellipsoid. Ovary 3–3.5 × 1–1.5 mm. narrowly ellipsoid, compressed, narrowed into a short stout style; stigma bilobed, papillate. Capsule 8 × 2 mm. long, narrowly ellipsoid. Seeds 0.35–0.4 mm. in diam., subglobose; testa faveolate.

Zimbabwe. E: Mutare Distr., Himalaya Mts., fl. & fr. 3.iii.1954, *Wild* 4478 (K; LISC).
Also Ethiopia. In damp short grassland; 2130 m.

12. **Swertia intermixta** A. Rich., Tent. Fl. Abyss. **2**: 57. —Baker & N.E. Br. in F.T.A. **4**, 1: 583 (1904). Type from Ethiopia.

A small annual herb up to 15 cm. tall. Stem slender, branching upwards, 4-ridged. Leaves in distant pairs, widening upwards, the upper 7–12 × 4–7 mm., obtuse at the apex; the lower smaller oblong-elliptic, narrowed into a very short petiole. Flowers 4-merous, white or cream yellow in terminal subcorymbiform cymes, pedicellate; pedicels very variable. Calyx almost divided to the base; lobes subequal, 3.5–5 × 0.9–1.2 mm., subspathulate, obtuse or subobtuse at the apex. Corolla with a short tube; lobes 5–5.5 × 1–1.5 mm., elliptic, obtuse at the apex with a pair of fimbriate nectaries within, near the base. Stamens inserted at the sinuses; filaments 1.5–3 mm. long; anthers 0.5 × 0.4 mm. Ovary 3–3.5 × 2–2.5 mm., broadly ellipsoid, compressed, subobtuse, narrowed into a short stout style; stigma, bilobed; lobes broad, obtusely rounded, papillate. Capsule 4–5 × 3.5–4 mm., broadly ellipsoid. Seeds 0.25–0.4 mm. in diam.; testa faveolate.

Zambia. N: Mbala Distr., Kawimbe, fl. & fr. 23.iii.1957, *Richards* 8841 (K). **Malawi**. S: Ntcheu Distr., Lower Kirk Range, Chipusiri, fl. & fr. 17.iii.1955, *Exell, Mendonça & Wild* 934 (BM; LISC). Also in Ethiopia, wet grassland among short grass; 1460–1680 m.

6. CHIRONIA L.

Chironia L., Sp. Pl.: 189 (1753).

Annual or perennial herbs, rarely suffrutescent. Stems erect, simple or branched at the apex or straggling, rarely radicant. Leaves cauline and in basal rosette (sometimes soon deciduous); the cauline ones linear, linear-lanceolate, ovate-lanceolate to ovate-cordate, sometimes much reduced, sessile, or amplexicaul, rarely cuneate to a petiole-like base. Flowers 5-merous in a lax, few–many-flowered cyme, sometimes solitary. Calyx 5-lobed, sometimes winged on the keel and decurrent on the tube, without basal glands. Corolla tube campanulate or cylindric; lobes 5, spreading, usually as long as or longer than the tube, without nectaries. Stamens 5, inserted at the throat of the corolla; filaments without scales at the base; anthers straight or twisted. Ovary globose to cylindric-globose, unilocular; ovules numerous; style simple, terete usually becoming hooked at the apex; stigma subcapitate or bilobed. Capsule septicidally bivalved, rarely baccate. Seeds numerous, subglobose; testa faveolate.

A relatively small genus with about 30 species throughout tropical and subtropical Africa and Madagascar.

1. Perennial herbs with a creeping rhizome; many leaves clustered at the base, persisting; flowers fastigiate, in a terminal cymose panicle (thyrsoid)   -   -   -   -   -   - 1. *krebsii*
– Annual or perennial herbs with a rootstock; some basal leaves subrosulate, sometimes not persisting; flowers in a rather lax cyme, not fastigiate   -   -   -   -   -   - 2
2. Stem always laxly branched; lower branches of the inflorescence up to 12 cm. long; cauline leaves ovate-cordate, ovate-triangular or triangular-lanceolate, 18–30 (35) mm. broad   -   -   -   -   -   -   -   -   -   -   -   -   -   - 3
– Stem unbrached or laxly branched; lower branches of the inflorescence up to 6 cm. long; cauline leaves elliptic, lanceolate, linear-lanceolate to linear 0.5–15 mm. broad   -   -   - 4
3. Inflorescence of open 2–3-flowered cymes; corolla tube 5–8 mm. long, with 18–25 mm. long lobes; capsule 18 × 5 mm.   -   -   -   -   -   -   -   -   - 6. *gratissima*
– Inflorescence of 2–7-flowered cymes arranged in a laxly terminal panicle; corolla tube 4–5 mm. long, with 10–18 mm. long lobes; capsule 5–6 × 2.5–3 mm.   -   -   -   - 7. *laxiflora*
4. Perennial or biennial herbs; pedicel up to 6–7 cm. long; calyx lobes 4–12 mm. long, ovate-lanceolate to linear, acuminate   -   -   -   -   -   -   -   -   -   -   -   - 5
– Annual herbs; pedicel up to 2.5 cm. long; calyx lobes 3–4.5 mm. long, triangular, acute   -   -   -   -   -   -   -   -   -   -   -   -   -   -   -   -   - 6
5. Calyx about as long as the corolla tube; bracts and calyx lobes not attenuate into a slender point; corolla tube 8–10 mm. long; corolla lobes 12–18 × 5–9 mm., oblong-lanceolate   -   -   -   -   -   -   -   - 2. *palustris* subsp. *transvaalensis*
– Calyx longer than the corolla tube; bracts and calyx lobes attenuate into a long point; corolla tube 4–5 mm. long; corolla lobes 8–12 × 3–5 mm., oblanceolate or obovate   -   -   -   -   -   -   -   -   -   - 3. *purpurascens* subsp. *humilis*
6. Cauline leaves linear-lanceolate, 0.5–2.5 mm. broad; corolla lobes oblong-ovate, 5–6 mm. long; anthers 1–1.5 mm. long   -   -   -   -   -   -   -   -   - 4. *flexuosa*
– Cauline leaves lanceolate or oblong-lanceolate, 3–15 mm. broad; corolla lobes lanceolate, 8–15 mm. long; anthers 3.5–5 mm. long   -   -   -   -   -   - 5. *elgonensis*

1. **Chironia krebsii** Griseb., Gen. et Sp. Gent.: 98 (1839). —Baker & N.E. Br. in F.T.A. **4**, 1: 554 (1903). —Hill & Prain in F.C. **4**, 1: 1107 (1909). —Marais & Verdoorn in Fl. S. Afr. **26**: 227 (1963). —Ross in Bot. Surv. Mem. **39**: 279 (1972). TAB. **11**. Type from S. Africa.

Tab. 11. CHIRONIA KREBSII. 1, habit (× ½); 2, flower-bud (× 2); 3, flower (× 2); 4, diagram of longitudinal section through corolla tube showing pistil and one stamen (× 2); 5, anther after dehiscence (× 4); 6, seed (× 15), all from *Chase* 1833.

Chironia densiflora Scott Elliot in Lond. Journ. of Bot. **29**: 69 (1891). —Engl., Pflanzenw.
Ost-Afr. **C**: 314 (1895). —Gilg in Engl. & Prantl, Nat. Pflanzenfam. IV, **2**: 77 (1892). Syntypes
from S. Africa (Natal and Cape).
Chironia palustris sensu Knobl. in Bot. Zentralbl. **60**: 329 (1894) pro parte, non Burch. (1824).

Perennial herb 40–70 cm. tall, sometimes with a creeping rhizome. Stem erect,
unbranched. Leaves mostly clustered at the base, often with 1–3 cauline pairs; the basal
ones 9–30 × 0.6–2 cm., oblanceolate to linear-spathulate gradually cuneate to the base; the
cauline ones 2.5–14 × 0.1–0.7 cm., usually linear to linear-lanceolate, acute at the apex.
Flowers magenta, rose-pink to red, rarely white, in a terminal cymose panicle; pedicels
1–3 cm. long; bracts 12–15 × 1–2.5 mm., oblong-ovate, slightly cucullate at the base, acute
at the apex. Calyx tube very short 1–2 mm. long; lobes 5–7 × 1.5–2 mm., linear-lanceolate,
acute at the apex, carinate, with a hyaline margin. Corolla tube 7–10 mm. long, cylindric;
lobes 5–13 (15) × 5–6 mm., oblong-lanceolate, narrowing towards the acute apex. Stamens
inserted about 1.5 mm. below the throat of the corolla; filaments 3.5–4 mm. long; anthers
c. 5 mm. long, ellipsoid, slightly spirally twisted. Ovary 7–8 × 1.5–2 mm., narrowly ellipsoid,
narrowing into a style c. 1 cm. long; stigma bilobed; lobes c. 2 mm. long. Capsule 9–12 ×
3–5 mm., narrowly ovoid. Seeds c. 0.35 mm. in diam., subglobose; testa faveolate.

Botswana. SE: Mochudi, fl. i.iv.1914, *Harbor* 6602 (K). **Zimbabwe**. E: Nyanga Distr., Inyangani
Mt., fl. 22.xi.1949, *Chase* 1833 (K; SRGH). **Malawi**. S: Zomba, fl. & fr. xii.1896, *Whyte* s.n. (K).
Mozambique. MS: Manica, Zuira Mt., Rd. to Chimoio (Vila Perry), fl. 12.xi.1965, *Torre & Pereira*
12937 (LISC).
Also in S. Africa. Growing in marshy ground; usually at high altitudes.

2. **Chironia palustris** Burch., Trav. **2**: 226 (1824). Type from S. Africa.

Subsp. **transvaalensis** (Gilg) Verdoorn in Bothalia **7**: 460 (1961). —Marais & Verdoorn in Fl. S. Afr.
26: 230, fig. 34, 2 (1963). —Friedrich-Holzhammer in Merxm., Prodr. Fl. SW. Afr. **110**,
Gentianaceae: 2 (1967). —Boutique in Fl. Afr. Centr., Gentianaceae: 34 (1972). Type from S.
Africa.
Chironia transvaalensis Gilg in Engl., Bot. Jahrb. **26**: 106 (1898). —Schoch in Bot. Zentrabl.,
Beih. **14**: 227 (1903). —Baker & N.E. Br. in F.T.A. **4**, 1: 555 (1903). —Hill & Prain in F.C. **4**, 1: 1105
(1909). —Mogg in Fl. Pl. Afr. **21**: t. 814 (1941). Type as above.

Erect annual or perennial herb up to 80 cm. tall with a woody rhizome.Stem
unbranched or branched above, 4-ridged. Basal leaves subrosulate, spathulate-ovate or
oblong-lanceolate, glaucescent, usually not persisting; cauline leaves 2–8 × 0.15–0.8 cm.,
much reduced and distant, linear-lanceolate to lanceolate or narrowly oblong-lanceolate,
narrowing to an acute apex, glaucous, not clearly 3-nerved. Flowers mauve, pink, rose-
pink, bright-pink to red, c. 3.5 cm. diam., in lax, pedicellate, 5–18 flowered cymes; pedicels
2–12 cm. long; bracts 5–20 × 1–3 mm., lanceolate. Calyx tube very short 1–2 mm. long;
lobes 4–5 × 1–2 mm., ovate-lanceolate acuminate, carinate, with hyaline margin. Corolla
tube 8–10 mm. long; lobes 12–18 × 5–9 mm., oblong-lanceolate, acute. Stamens inserted
about 1 mm. below the throat of the corolla; filaments 2–3 mm. long; anthers 4–7 mm.
long, ellipsoid, slightly spirally twisted, yellow. Ovary 6–7 × 2–3 mm., narrowly ellipsoid,
narrowing into a style c. 1.2 cm. long; stigma compressed, shortly lobed. Capsule 8–13 ×
4–6 mm., narrowly ovoid. Seeds c. 0.25–0.30 mm. in diam., testa faveolate.

Zambia. B: c. 8 km. from Senanga ferry, along road to Katima Mulilo, fl. 23.x.1972, *Strid* 2420
(K). N: Lake Kashiba, fl. 22.x.1957, *Fanshawe* 3813 (K). W: Ndola, fl. 7.ii.1960, *Robinson* 3367
(K). C: c. 8 km. E. of Lusaka, fl. 4.xii.1955, *King* 233 (K). S: Victoria Falls, fl. & fr. 1954, *Levy* 1136
(PRE). **Zimbabwe**. N: Makonde Distr., foot of Mutoroshanga Pass, fl. 4.ix.1960, *Leach & Bayliss* 10474
(K). W: Shangani, fl. ii.1943, *Feiertag* in GHS 45520 (K; PRE). C: Harare, fl. xi.1955, *Drummond* 4939
(K; LISC). **Mozambique**. MS: Zinhumbo Hills, fl. 23.xi.1949, *Swynnerton* 1891 (BM; K).
Also in S. Africa, Namibia and Angola. Growing in marshy grasslands on banks of streams and
rivers at high altitudes; 1000–3500 m.
Chironia palustris subsp. *transvaalensis* has the cauline leaves well developed and the basal ones
usually not persistent, whereas subsp. *palustris* (S. Africa: Transvaal, Orange Free State, Natal,
Basutoland and Cape Prov.) has cauline leaves much reduced and usually in 1–3 pairs and basal
leaves in a persistent rosette.
Subsp. *transvaalensis* is distinguishable from subsp. *rosacea* (Natal and Swaziland) by having the
cauline leaves usually up to 8 mm. broad, clearly 3-nerved and flowers under 25 mm. long, whereas
the latter has cauline leaves up to 30 mm. broad, the broad ones obviously 3-nerved and flowers
about 30 mm. long.

3. **Chironia purpurascens** (E. Mey.) Benth. & Hook. f., Gen. Pl. **2**: 805 (1876). Type from S. Africa.

Subsp. **humilis** (Gilg) Verdoorn in Bothalia **7**: 462 (1961). —Marais & Verdoorn in Fl. S. Afr. **26**: 232 , fig. 34, 1 (1963). —Ross in Bot. Surv. Mem. **39**: 279 (1972). Type from S. Africa (Transvaal).

    Chironia humilis Gilg in Engl., Bot. Jahrb. **26**: 105 (1898). —Baker & N.E. Br. in F.T.A. **4**, 1: 555 (1903). —Hill & Prain in F.C. **4**, 1: 1107 (1909). —Dyer in Fl. Pl. Afr. **24**: t. 95 (1944). Type as above.

    Chironia wilmsii Gilg in Engl., Bot. Jahrb. **26**: 105 (1898). Type as above.

    Chironia humilis var. *wilmsii* (Gilg) Prain in Kew Bull. **1908**: 350 (1908). —Hill & Prain, tom. cit.: 1107 (1909). Type as above.

    Chironia humilis var. *zuluensis* Prain in Kew Bull. **1910**: 55 (1910). Type as above.

Biennial or perennial herb up to 50 cm. tall, with a slender rhizome. Stem slender, unbranched or cymosely at the apex, 4-ridged. Leaves sub-rosulate, at the base, often with 4 or more cauline pairs; the rosette ones 0.7–6 × 0.1–1 cm., lanceolate, linear or linear-lanceolate, subacute at the apex soon disappearing; cauline ones, 1.5–6 × 0.1–1 cm., narrowly linear to elliptic or oblong-elliptic, usually narrowed to an acute or obtuse apex. Flower pure white, pink-mauve or bright pink c. 2 cm. in diam., in a fairly lax cyme; the central flower of cyme sessile or with a pedicel up to 6 mm. long, bracteate; bracts 5–12 × 1–2 mm., attenuate into a long point. Calyx tube 1–2 mm. long; lobes 7.5–12 × 1.8–2.5 mm., linear lanceolate, acuminate and attenuate into a long point, hyaline margined and carinate. Corolla tube 4–5 mm. long; lobes 8–12 × 3–5 mm., oblanceolate or obovate, acute. Stamens inserted about 1.5 mm. below the throat of the corolla; filaments 3–4 mm. long; anthers 2–3 mm. long, strongly twisted. Ovary 3.5–5 × 1–1.5 mm., narrowly ellipsoid, narrowing into a style c. 6 mm. long; stigma 0.5–0.7 mm. long, bilobed. Capsule 5–7 × 2.5–3.5 mm., oblong-ovoid. Seeds c. 0.25 mm. in diam.; testa faveolate.

    Zimbabwe. W: Matopos, fl. 16.ii.1941, *Hopkins* in GHS 7979 (K). C: Rusape, fl. & fr. 5.ii.1949, *Munch* 160 (K; LISC). E: Nyanga (Inyanga), fl. 1.i.1973, *Biegel* 4134 (K; LISC; PRE; SRGH). S: Masvingo, fl. & fr. 1909–12, *Monro* 1721 (BM). **Mozambique**. M: Namaacha, fl. & fr. 27.iii.1957, *Barbosa & Lemos* 7541 (COI; K; LISC).

    Also in S. Africa. Growing in damp situations, near marshy places on mountain tops; 1850–1925 m. *Chironia purpurascens* subsp. *purpurascens* (Swaziland, Natal and Cape Prov.) is 50–80 cm. tall and has the pedicels of the central flower in each cyme 6–25 mm. long, or longer, whereas subsp. *humilis* is smaller (up to 50 cm. tall) and has the central flower of each cyme sessile or with a pedicel up to 6 mm. long.

4. **Chironia flexuosa** Baker in Kew Bull. **1908**: 296 (1908). —Boutique in Fl. Afr. Centr., Gentianaceae: 36 (1972). Type: Zambia, Kambole, SW. of Lake Tanganyika, 1896 *Nutt* s.n. (K; holotype).

Annual, erect herb up to 80 cm. tall. Stem unbranched above, 4-ridged. Basal leaves subrosulate; lamina 20 × 10 mm., oblong-ovate, obtuse at the apex, narrowed to the base; the cauline ones, 8–50 × 0.5–2.5 mm., sessile, linear-lanceolate, c. 8 pairs, 1-nerved. Flowers pedicellate, pink mauve or bright red in 1–3-flowered cymes, arranged in terminal lax panicles up to 18 cm. long, bracteate; bracts 3–8 × 0.8–1 mm.; pedicels 0.5–2.5 cm. long. Calyx tube 1.5–2 mm. long; lobes 3–3.5 mm. long, triangular, acute. Corolla tube 4–5 mm. long; lobes 5–6 × 2.5 mm., oblong-ovate, subobtuse or subacute at the apex. Stamens inserted about 1 mm. below the throat of the corolla; filaments 2 mm. long; anthers 1–1.5 mm. long, pallid yellow, slightly spirally twisted. Ovary 5 mm. long, cylindric, narrowing into a style c. 3 mm. long; stigma bilobed. Capsule 6–7 × 2.5–3 mm., cylindric to cylindric-ovoid. Seeds 0.3 mm. in diam.; testa faveolate.

    Zambia. N: Kambole, SW. of Lake Tanganyika, fl. & fr. 1896, *Nutt* s.n. (K). W: Mufulira, fl. & fr. 14.vi.1934, *Eyles* 8166 (BM; K). C: Chakwenga, Headwaters, 100–129 km. E. of Lusaka, fl. 27.iii.1965, *Robinson* 6479 (K).

    Also in Zaire. In swampy, river banks, moist grasslands; 1200–1500 m.

5. **Chironia elgonensis** Bullock in Kew Bull. **1932**: 500 (1932). —Boutique in Fl. Afr. Centr., Gentianaceae: 37 (1972). Type from Kenya (Mt. Elgon).

Annual erect herb up to 120 cm. tall. Stem unbranched or laxly branched above, 4-ridged. Stem leaves 10–40 × 3–15 mm., lanceolate or oblong-lanceolate, subacute at the apex, semi-amplexicaul at the base. Flowers pedicellate, crimson, rose, strawberry pink, or red, in 1–2 (3) flowered cymes arranged in terminal lax panicles, bracteate; bracts 2–10 × 0.8–1 mm.; pedicels 0.7–2 cm. long. Calyx tube 1–1.5 mm. long; lobes 3–4.5 mm. long, triangular, acute at the apex. Corolla tube 3–4.5 mm. long; lobes 8–15 × 3–4 mm., lanceolate, acute at the apex. Stamens inserted about 1 mm. below the throat of the corolla; filaments 1.5–2 mm. long; anthers 3.5–5 mm. long, linear, spirally twisted, yellow. Ovary 6 × 3 mm., narrowly ellipsoid, narrowing into a style 4–8 mm. long; stigma bilobed, claviform. Capsule 6–8 × 3.5–4 mm. cylindric-ovoid. Seeds 0.15–0.2 mm. in diam.; testa faveolate.

Zambia. N: Mbala Distr., Kalambo Farm, Saisi Valley, fl. & fr. 21.v.1952, *Richards* 1799 (K). Also in Burundi, Kenya and Tanzania. In wet marshes among tall grass; 1800 m.

6. **Chironia gratissima** S. Moore in Journ. Linn. Soc., Bot. **40**: 148 (1911). —Boutique in Fl. Afr. Centr., Gentianaceae: 37 (1972). Type: Zimbabwe, Chimanimani, *Swynnerton* 1892 (K, holotype).

Laxly trailing or scrambling annual or perennial herb up to 1.50 m. tall. Stem branched, 4-ridged. Cauline leaves (3) 6–9 (10) × (1) 2.5–3 cm., ovate-triangular-lanceolate, acute at the apex, semi-amplexicaul at the base, conspicuously 3-nerved, pale yellow green. Flowers mauve, bright mauve, or deep magenta pink or purple in 2–3 flowered cymes arranged in terminal lax panicles, bracteate; bracts 3–10 mm. long; pedicels 1–3.5 cm. long. Calyx tube c. 2 mm. long; lobes 6–8 × 1–1.5 mm. dorsally carinate. Corolla tube 5–8 mm. long; lobes 18–25 × 3–7 mm., oblong-lanceolate, apiculate at the apex. Stamens inserted about 1 mm. below the throat of the corolla; filaments 1.5–3.5 mm. long; anthers 5–8 mm. ellipsoid, yellow or pale yellow, twisted. Ovary 8–10 × 4–5 mm. narrowly-ellipsoid, narrowing into a style c. 10 mm. long; stigma bilobed c. 2 mm. long, oblong. Capsule 18 × 5 mm., cylindric-ovoid. Seeds 0.25–0.3 mm. in diam.; testa faveolate.

Zimbabwe. E: Chimanimani, fl. 10.x.1960, *Wild* 356A (K; LISC). **Mozambique**. MS: Manica, Zuira Mt., Tsetserra road to Chimoio, 2100 m., fl. & fr. 2.iv.1966, *Torre & Correia* 15589 (LISC). Also in Zaire. On rocky or stream banks and in damp soil; 1160–2100 m.

7. **Chironia laxiflora** Baker in Kew Bull. **1894**: 25 (1894). —Baker & N.E. Br., in F.T.A. **4**, 1: 556 (1903). Type: Malawi, Manganja Hills, *Meller* s.n. (K, holotype).
    Chironia rubrocaerulea Gilg in Engl., Bot. Jahrb. **30**: 379 (1902). Type from Tanzania.

Spreading or weakly erect annual herb, 30–150 cm. tall. Stem slender branched, 4-ridged. Stem leaves in distant pairs, 1.2–4 (5) × 1.8–2.5 (3.5) cm., ovate-cordate, acute to acuminate at the apex, amplexicaul at the base, 3–5 nerved. Flowers pedicellate, star-like, crimson, pink, pinkish-purple, deep magenta to red, in 2–7 flowered cymes arranged in a laxly terminal panicle, bracteate; bracts 3–13 mm. long, lanceolate; pedicels 0.8–3 cm. long. Calyx tube 1–2 mm. long; lobes 5–8 × 1–1.5 mm. long, lanceolate-acuminate. Corolla tube 4–5 mm. long; lobes 10–18 × 3–4 mm., lanceolate, acute at the apex. Stamens inserted about 1 mm. below the throat of the corolla; filaments 1.5–2 mm. long; anthers yellow or orange-yellow 8 mm. long, lanceolate, spirally twisted. Ovary 3–5 × 4 mm., narrowly-ovoid, narrowing into a style c. 4–5 mm. long; stigma bilobed, 2 mm. long. Capsule 5–6 × 2.5–4 mm., cylindric-ovoid. Seeds c. 0.25–0.3 mm. in diam.; testa faveolate.

Zambia. N: Lunzua River c. 20 km. W. of Mbala, fl. & fr. 14.v.1951, *Bullock* 3973 (K). **Malawi**. N: Chitipa Distr., Misuku Hills, Mughesse, fl. 25.iv.1972, *Pawek* 5193 (BM). C: Lilongwe Distr., Dzalanyama Forest Reserve, above Chaulongwe Falls, fl. & fr. 26.iv.1970, *Brummitt* 10168 (K). S: Blantyre Distr., Upper Hynde Dam, 2 km. N. of Limbe, fl. & fr. 14.ii.1970, *Brummitt* 8551 (K). **Mozambique**. N: Lichinga, fl. & fr. v.1934, *Torre* 91 (LISC). Also in Tanzania. On damp sandy loam, open places in miombo woodland and in partial shade of riverine forest; 1300–1500 m.

7. CANSCORA Lam.

Canscora Lam., Encycl. Meth., Bot. **1**: 601 (1785). —Benth. in Benth. & Hook. f. Gen. Pl. **2**: 811 (1876).

Annual herbs. Stems erect, much branched, 4-ridged or narrowly 4-winged, glabrous. Leaves sessile, sometimes connate and decurrent on a ridge or narrowly winged; lamina lanceolate, elliptic, ovate to ovate-lanceolate, 3–5-nerved. Flowers 4-merous, in terminal or axillary, lax cymes, sometimes, solitary. Calyx tube terete or winged; lobes much shorter than the tube. Corolla regular or rarely irregular; tube cylindric, with 4 equal or unequal (2 large and 2 short) lobes, much shorter than the tube. Stamens 4, inserted at different levels in the corolla tube, often some sterile; filaments short; anthers ovoid or ellipsoid not twisted. Ovary ellipsoid or cylindric, unilocular; ovules numerous; style filiform; stigma bilobed. Capsule cylindrical, bivalved. Seeds numerous; testa minutely faveolate.

A tropical genus with 12 species from Indonesia, Malaysia, Australia and Africa.

1. Stems, pedicels and calyx tube 4-winged; corolla tube 6–8 mm. long; lobes 3–4 mm.
   long - - - - - - - - - - - - - - 1. *decussata*
 – Stems, pedicels and calyx tube not winged; corolla tube 2.5–6 mm. long; lobes 1.5–2 mm.
   long - - - - - - - - - - - - - - 2
2. Cymes laxly diffuse; pedicels filiform, 5–15 mm. long; calyx teeth subulate, 1.5–2 mm. long;
   capsule 4–5 mm. long - - - - - - - - - - 2. *diffusa*
 – Cymes slender; pedicels 4–12 mm. long; calyx teeth deltoid acute, 0.75–1.5 mm. long; capsule
   3–3.5 mm. long - - - - - - - - - - - 3. *kirkii*

1. **Canscora decussata** (Roxb.) Roem. & Schult. in Syst. Veg. **3**: 229 (1827). —Hook. in Curtis, Bot. Mag., N. Ser., **5**: t. 3066 (1831). —Baker & N.E. Br. in F.T.A. **4**, 1: 557 (1903). —Hutch. & Dalz. in F.W.T.A. **2**, 1: 183, fig. 235 A–F (1931). —Taylor in F.W.T.A., ed. 2, **2**: 300, fig. 268 A–F (1963). —Boutique in Fl. Afr. Centr., Gentianaceae: 30, fig. 2A (1972). Type from India.
   Pladera decussata Roxb., Fl. Ind. **1**: 418 (1820). Type as above.
   Canscora diffusa Auct. —De Wild. in Pl. Bequaert. **2**: 110 (1923) pro parte, non R. Br. ex Roem. & Schult.

Annual erect herb, 15–30 cm. tall, much branched; stem conspicuously 4-winged, glabrous. Leaves sessile or narrowed into a petiole-like base; lamina 7–25(30) × 2–10(15) mm., lanceolate, ovate to narrowly ovate-lanceolate, acute at the apex, obtuse at the base, 3-nerved from the base, glabrous. Flowers white or pale pink, in terminal and axillary lax cymes, rarely solitary; bracts 3–8 × 1–2 mm., ovate-lanceolate, sessile, 3-nerved; pedicels 3–10 mm. long, stout, 4-winged. Calyx tube 5–8 mm. long, swollen towards the base, 4-winged; teeth ovate-lanceolate, 1.5 mm. long. Corolla tube 6–8 mm. long; lobes 3–4 × 2.5–3.5 mm., unequal, obovate, obtuse or emarginate. Filaments 0.2–0.5 mm. long; anthers c. 1 mm. long, ovoid-ellipsoid; sterile stamens 1–2. Ovary 5–6 mm. long, cylindric-ellipsoid; style c. 3 mm. long, linear; stigmatic lobes c. 2 × 2 mm., ovoid-globular. Capsule 5.5–7.5 mm. long, cylindric. Seeds numerous, c. 0.3 mm. in diam.; testa minutely faveolate.

Malawi. N: Khondowe (Kondowe) to Karonga, fl. & fr. vii.1896, *Whyte* 81 (K). **Mozambique**. Z: Namagoa, Mocuba, fl. & fr. *Faulkner* 304 (COI; P; PRE; SRGH).
Widespread throughout tropical Africa and Madagascar; also in tropical Asia and Australia. Damp places and marshy ground.

2. **Canscora diffusa** (Vahl) R. Br. ex Roem. & Schult. in Syst. Veg. **3**: 301 (1820). —Baker & N.E. Br. in F.T.A. **4**, 1: 558 (1903). —Hutch. & Dalz. in F.W.T.A., **2**: 183 (1931). —Taylor in F.W.T.A., ed. 2, **2**: 300 (1963). —Boutique in Fl. Afr. Centr., Gentianaceae: 32, fig. 2B (1972). Type from Egypt.
   Gentiana diffusa Vahl in Symb. Bot. **3**: 47 (1794). Type as above.

Annual erect herb, up to 35 (70) cm. tall, much branched. Stem slender, 4-angled, glabrous. Upper leaves sessile, the lower shortly petiolate; petiole up to 5 mm. long; lamina 1–3 (4) × 0.5–1.5 cm., ovate-elliptic to elliptic, acute at the apex, largely cuneate at the base, 3–5 nerved from the base, glabrous. Flowers very numerous, pink, mauve or white, in lax somewhat diffuse, leafy cymes; bracts 4–8 × 2–5 mm., sessile, ovate, acute; pedicels 5–15 mm. long, filiform, not winged. Calyx tube 4–5 mm. long, narrow, scarcely swollen towards the base, not winged; teeth subulate, 1.5–2 mm. long. Corolla tube 5–6 mm. long; lobes 1.5–2 × 1–1.5 mm., unequal, ovate obtuse. Filaments 0.4–0.9 mm. long; anthers 0.25–0.35 × 0.25 mm., ellipsoid; sterile stamens 1–2. Ovary 0.3–0.4 mm. long, cylindric; style 1–1.5 mm. long, linear; stigmatic lobes, 0.4 mm. long. Capsule 4–5 mm. long, cylindric. Seeds, 0.2 mm. in diam.; numerous; testa minutely faveolate.

Zambia. W: N. of Chingola, fl. & fr. 4.v.1960, *Robinson* 3702 (K). C: Lusaka, fl. & fr. 29.v.1958, *Fanshawe* 4462 (K). E: Chipata, fl. & fr. 13.x.1967, *Mutimushi* 2334 (LISC; PRE; SRGH). S: c. 47 km. N. of Choma, fl. & fr. 29.v.1954, *Robinson* 769 (K). **Zimbabwe**. N: Gokwe, Sengwa Research St., fl. & fr. 17.x.1968, *Jacobsen* 252 (K; PRE; SRGH). E: Mutare, fl. & fr. 27.viii.1959, *Phipps* 2172 (K; SRGH). **Malawi**. N: Chitipa, Songa Stream, 14.5 km. E. of crossroad on Rd. to Karonga, fl. & fr. 8.vii.1973, *Pawek* 7169 (K). S: Nsanje Distr., Mwabvi Game Reserve, fl. & fr. 6.viii.1975, *Salubeni* 1982 (PRE; SRGH). **Mozambique**. N: Mutuali, Cucuteia Mt., fl. & fr. 16.iii.1964, *Torre & Paiva* 11218 (LISC). Z:

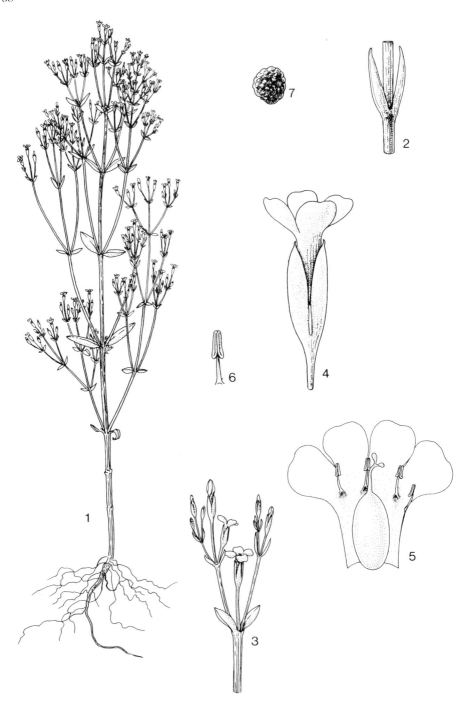

Tab. 12. CANSCORA KIRKII. 1, habit ($\times\frac{1}{2}$); 2, stem node showing opposite leaves ($\times$ 6); 3, part of inflorescence ($\times$ 2); 4, flower ($\times$ 6); 5, corolla opened out showing pistil and stamens ($\times$ 6); 6, anther ($\times$ 12); 7, seed ($\times$ 40), all from *Fanshawe* 6997.

Namagoa Distr., Lugela, Mocuba, fl. & fr. vii.1943, *Faulkner* 99 (COI; K; LISC; PRE). T: Tete, Cahobra Bassa, Mucangadzi R., fl. & fr. 19.v.1972, *Pereira & Correia* 2724 (LISC; LMU). MS: Gorongosa, fl. & fr. 1884–85, *R. Carvalho* s.n. (COI).
    Widespread in tropical Africa; also in tropical Asia and Australia. Growing on rock surfaces damp sands and river banks; 610–1350 m.

3. **Canscora kirkii** N.E. Br. in F.T.A. **4**, 1: 558 (1903). TAB. **12**. Type: Zimbabwe, Is. in the Zambezi R. at Victoria Falls, *Kirk* s.n. (K, holotype).

Annual erect herb, up to 30 (35) cm. tall. Stem slender, 4-angled, glabrous. Upper leaves sessile, the lower shortly petiolate; petiole up to 5 mm. long; lamina 1–3.5 × 0.5–1.5 (2.5) cm., ovate-elliptic to elliptic, acute at the apex, widely cuneate at the base, 3–5 nerved from the base, glabrous. Flowers very numerous, pink, pinkish-white or white, in slender leafy cymes; bracts 3–5 × 1.5–3 mm., sessile, ovate, acute; pedicels 4–12 mm. long, slender, not winged. Calyx tube 2–2.5 mm. long; teeth deltoid acute, 0.75–1.5 mm. long. Corolla tube 2.5–3.5 mm. long; lobes 1.5–2 × 1–1.5 mm., obovate-elliptic, oblong, emarginate or bifid at the apex. Stamens all alike and fertile; filaments 0.5–0.7 mm. long; anthers 0.5 mm. long. Ovary 0.2–0.3 mm. long, cylindric; style c. 1 mm. long, linear; stigmatic lobes c. 0.3 mm. long, obovate. Capsule 3–3.5 mm. long, cylindric. Seeds 0.2–0.35 mm. diam., subglobose, numerous, minutely scrobiculate.

    Zambia. S: Victoria Falls, fl. & fr. 7.vii.1930, *Hutchinson & Gillett* 3439 (K; BM). **Zimbabwe**. W: Is. in the Zambezi R. at Victoria Falls, *Kirk* s.n. (K).
    So far known only from the Victoria Falls, bog at edge of rain forest and rocks in spray zone.

8. SCHINZIELLA Gilg

Schinziella Gilg in Engl. & Prantl, Nat. Pflanzenfam. 4, **2**: 74 (1892).

Erect perennial herbs. Stems not or sparsely branched from the base. Leaves sessile, the upper ones much reduced and scale-like. Flowers 4-merous in compact, pedunculate, terminal heads. Calyx tube as long as the lobes. Corolla regular with a cylindric tube and with 4 lobes about as long as the tube, without nectaries. Stamens 4, unequal, 3 sterile, 1 larger, fertile. Ovary cylindric-ovoid, unilocular; ovules numerous; style filiform; stigma bilobed. Capsule narrowly-obovoid, bivalved. Seeds numerous, subglobose.

A monotypic genus from tropical Africa.

Schinziella tetragona (Schinz) Gilg in Engl. & Prantl, Nat. Pflanzenfam. 4, **2**: 74 (1892). —Baker & N.E. Br. in F.T.A. **4**, 1: 557 (1903). —Taylor in F.W.T.A. ed. 2, **2**: 300 (1963). —Boutique in Fl. Afr. Centr., Gentianaceae: 29 (1972). TAB. **13**. Type from Angola.
    Canscora tetragona Schinz in Vierteljahrschr. Nat. Ges. Zürich **37**: 388 (1891). Type as above.
    Schinziella tetragona var. *parviflora* Schinz ex De Wild. in Pl. Bequaert. **5**: 421 (1932). Syntypes from Zaire.

Herb 15–60 cm. tall, not or very sparsely branched from the base. Stem 4-angled, broadly winged. Leaves sessile, the lowermost 5–15 × 3–10 mm., elliptic-ovate to elliptic obovate, obtuse or shortly apiculate at the apex, subcoriaceous; the uppermost 2–5 × 1–2.5 mm., scale-like, linear to narrowly lanceolate, acuminate at the apex. Flowers white or yellowish, soon turning brown, sessile or subsessile (pedicels 0.3–2 mm. long), numerous in compact terminal head-like, pedunculate, cymes 10–20 mm. in diam.; peduncle 0.2–5 mm. long; bracts 2–3 mm. long, linear-lanceolate, acuminate; bracteoles 2–5 mm. long, linear-lanceolate, acute. Calyx tube 2–2.3 mm. long; lobes 2–2.25 × 1.2–2 mm., triangular acuminate at the apex, 2–3-nerved. Corolla tube 2.5–3 mm. long; lobes 2.5–3 × 1.8–2.5 mm., oblong-ovate, acute at the apex. Filament of fertile stamen 1–1.5 mm. long; anther c. 0.5 mm.; sterile filaments 0.2 mm. long; anthers 0.6–0.7 mm. long, subdeltoid. Ovary 1–1.5 × 0.6 mm., cylindric-ellipsoid; style 1–1.5 mm. long; stigma bilobed. Capsule 2.5–4 × 1.5–2 mm. cylindric-ellipsoid. Seeds c. 0.4 mm. in diam., prismatic-subconical; testa scrobiculate.

    Zambia. N: 72 km. S. of Mbala on Kasama Rd., fl. & fr. 30.ii.1955, *Exell, Mendonça & Wild* 1388 (BM; LISC). **Malawi**. C: Nkhota Kota, fl. & fr. 16.ii.1944, *Benson* 28019 (PRE). **Mozambique**. N: Niassa, c. 35 km. from Marrupa to Mecula, fl. & fr. 13.viii.1981, *Jansen, de Koning & Wilde* 277 (K).
    Widespread throughout tropical Africa. Wet places; c. 560 m.

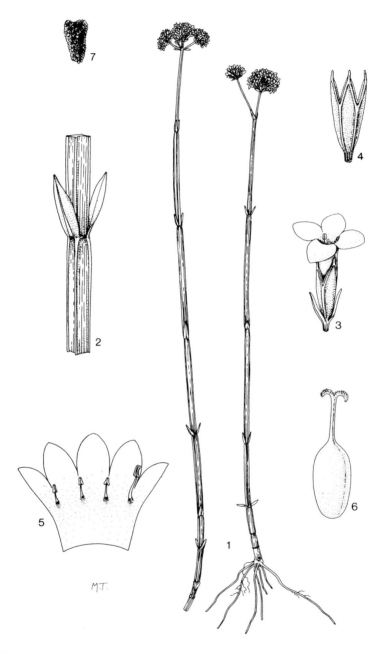

Tab. 13. SCHINZIELLA TETRAGONA. 1, habit (×½); 2, detail of stem, showing angles, node and leaf pair (× 2); 3, flower (× 4); 4, calyx (× 4); 5, corolla opened out showing 3 staminodes and 1 fertile stamen (× 4); 6, pistil (× 6); 7, seed (× 20), all from *Exell, Mendonça & Wild* 1338.

9. FAROA Welw.

Faroa Welw. in Trans. Linn. Soc., Bot. **27**: 45 (1869).

Annual or perennial herbs, sometimes somewhat succose. Stems erect to prostrate, simple or more or less branched, terete to 4-angled, usually more or less narrowly winged. Leaves linear to ovate, often connate, sometimes dilated and scarious at the base. Flowers 4-merous, usually pedicellate, in axillary few to many-flowered fascicles. Calyx a campanulate tube with 4 erect lobes. Corolla tube equaling or shorter than the calyx, with 4 more or less semilunate, fimbriate-papillose scales, subtending the filaments; lobes 4, spreading, ovate-lanceolate, as long as or shorter than the tube. Filaments inserted in the sinuses of the corolla lobes, filiform; anthers ovoid. Ovary more or less obovoid, unilocular; ovules numerous; style filiform; stigma simple, subcapitate, or bilobed, or obsolete, or divided in 2 filiform lobes. Capsule obovoid, bivalved. Seeds numerous, subglobose, minutely scrobiculate.

A fairly small genus of about 17 species throughout tropical Africa.

1. Acaulescent herb with a rosette of prostrate leaves; inflorescence a solitary cluster of flowers, sessile in the centre of the rosette of leaves - - - - - - - - - - 2
 - Herb with a well developed stem or stems, with or without a basal rosette of leaves; inflorescences usually of several clusters of flowers disposed along a more or less elongated axis - - - - - - - - - - - - - - - - - - - 3
2. Corolla lobes 3–4 mm. long, usually shorter than the filaments; style 5–7 mm. long - - - - - - - - - - - - - - - - - - 1. *duvigneaudii*
 - Corolla lobes 1–1.5 mm. long, usually longer than the stamen filaments; style (1.3) 1.5–2.5 mm. long - - - - - - - - - - - - - - - - - 2. *acaulis*
3. Plant with a more or less well developed basal rosette of leaves, these longer than those subtending the flower fascicles - - - - - - - - - 3. *salutaris*
 - Plant without a basal rosette of leaves - - - - - - - - - - 4
4. Corolla lobes externally subapically corniculate - - - - - - 4. *corniculata*
 - Corolla lobes not corniculate - - - - - - - - - - - - 5
5. Calyx lobes 3 or more-nerved from the base with anastomosing secondary nerves near the apex - - - - - - - - - - - - - - - - - 6
 - Calyx lobes 1-nerved, often winged or carinate - - - - - - - - 7
6. Calyx lobes 3-nerved from the base; corolla tube 1.5–1.8 (2) mm. long, tube and lobes externally papillose - - - - - - - - - - - - - - 5. *axillaris*
 - Calyx lobes with a conspicuous midrib and numerous sub-parallel nerves from the base; corolla tube 2.5–3 mm. long, externally smooth and lobes internally minutely papillose - - - - - - - - - - - - - - - 6. *involucrata*
7. Corolla lobes 0.4–1.25 mm. long, longer than the filaments (0.25–0.5 mm.long) 8
 - Corolla lobes 2–3 mm. long, as long as or shorter than the filaments (1.7–4.5 mm. long) - - - - - - - - - - - - - - - - - 10
8. Leaves not or scarcely exceeding the subtended flower fascicle; corolla tube smooth - - - - - - - - - - - - - - - - - 7. *amara*
 - Leaves much exceeding the subtended flower fascicle; corolla tube smooth or papillose - - - - - - - - - - - - - - - - 9
9. Corolla tube smooth, lobes 0.4–0.6 × 0.35 mm.; leaves linear-lanceolate, 5–15 × 1–1.5 mm.; calyx lobes without a short erect apiculus - - - - - - 8. *minutiflora*
 - Corolla tube usually papillose, lobes 1–1.25 × 0.5–0.7 mm.; leaves oblong-elliptic, 5–35 × 1.5–15 mm.; calyx lobes with a short erect apiculus - - - - - 9. *pusilla*
10. Corolla tube 3–4.2 mm. long - - - - - - - - - - 10. *alata*
 - Corolla tube 1.2–2.5 mm. long - - - - - - - - - - - 11
11. Caespitose succose perennial; upper leaves not or scarcely exceeding the subtended flower fascicle; inflorescence congested, subcapitate - - - - - 11. *hutchinsonii*
 - Erect or decumbent annual, branched above the base; upper leaves usually exceeding the subtended flower fascicle; inflorescence elongated - - - - - 12
12. Calyx lobes more or less winged; pedicel not papillate; fascicles of 3–9 flowers - - - - - - - - - - - - - - 12. *fanshawei*
 - Calyx lobes not winged; pedicel papillate; fascicles of c. 20 flowers - - 13. *affinis*

1. **Faroa duvigneaudii** Lambinon in Bull. Soc. Roy. Bot. Belg. **91**: 177 (1959). —Boutique in Fl. Afr. Centr., Gentianaceae: 20 (1972). —Taylor in Garçia de Orta, Sér. Bot. **1**: 74, t. 6 fig. 1–6 (1973). Type from Zaire.

Small acaulescent herb, 6–10 mm. tall. Leaves forming a prostrate rosette, petiolate; petiole shorter than the lamina; lamina 4–65 × 2.3–30 mm., ovate deltate or ovate-oblong, cuneate to round at the base, obtuse at the apex, the lower smaller. Inflorescence a solitary dense hemispheric cluster, sessile in the centre of the leaf rosette. Flowers pink, 5–22 mm. in diam.; pedicels c. 0.5 mm. long. Calyx tube 1–1.5 mm. long; lobes 2–2.5 × 1 mm., oblong-elliptic. Corolla tube 3–4 mm. long; lobes 3–4 × 1.3–1.5 mm., ovate-lanceolate, acute. Filaments 3–4 mm. long, longer than the corolla lobes; anthers 0.8–1 mm. long, ellipsoid. Ovary 1–1.5 × 0.5–0.7 mm., cylindric-ovoid; style 5–7 mm. long; stigma subcapitate. Capsule 2–2.5 × 1–1.2 mm., cylindric-ovoid. Seeds c. 0.5 mm. in diam., subglobose, minutely scrobiculate.

Zambia. W: Mwinilunga, fl. & imat. fr. 13.v.1972, *Kornaś* 1754 (K). S: Mazabuka, fl. 23.vii.1952, *Angus* 10 (K).
Also in Zaire. Damp grassland.

2. **Faroa acaulis** R.E. Fries in Wiss. Ergebn. Schwed. Rhod.-Kongo-Exped. 1911–1912, 1: 260 (1916). —Boutique in Fl. Afr. Centr., Gentianaceae: 22 (1972). —Taylor in Garçia de Orta, Sér. Bot. 1: 74, t. 6 fig. 7–12 (1973). Type: Zambia, Ndola, *Fries* 501a (UPS, holotype).
    Faroa paradoxa Gilg ex De Wild., Ann. Mus. Congo, Bot., Ser. 5, 2: 337 (1908) nom. nud.
    Faroa pygmaea Mildbr. in Notizbl. Bot. Gart. Mus. Berl. 11: 403 (1932). Type from Tanzania.

Small acaulescent herb, 5–15 mm. tall. Leaves in a prostrate rosette of 3–4 pairs, petiolate; petiole shorter than the lamina; lamina 6–45 × 3–32 mm., ovate-deltoid or ovate, cuneate to round at the base, obtuse at the apex; the lower smaller. Inflorescence a dense solitary hemispheric cluster, sessile in the centre of the leaf rosette. Flowers violet, blue, rose to whitish, 5–15 mm. in diam.; pedicels 0.5–12 mm. long. Calyx tube 0.8–1 mm. long; lobes 2–3 × 0.5–0.8 mm. long, spathulate. Corolla tube 1.5–2.5 mm. long; lobes 1–1.5 × 0.6–0.8 mm., ovate-lanceolate, subacute. Filaments 0.5–1.3 mm. long, shorter than the corolla lobes; anthers 0.5–0.6 mm. long, ellipsoid. Ovary 0.1–1.5 × 0.5–0.7 mm., ellipsoid-cylindric; style (1.3)1.5–2.5 mm. long; stigma bilobed. Capsule 1.8–2 × 1–1.2 mm., ellipsoid-cylindric. Seeds c. 0.5 mm. in diam., minutely scrobiculate.

Zambia. N: Mbala Distr., Kambole Escarpment, fl. & fr. 4.vi.1957, *Richards* 9996 (K). W: Ndola, fl. & fr. 10.v.1954, *Fanshawe* 1182 (K). C: Near Mumbwa, fl. & fr. 1911, *Macaulay* 332 (K). **Malawi**. N: Chitipa Distr., Nthalire descent road from Nyika, fl. & fr. 26.iv.1973, *Pawek* 6613 B (K).
Also in Angola, Zaire, Rwanda, Burundi and Tanzania. Open dry or damp grassland often on sand; 700–1600 m.

3. **Faroa salutaris** Welw. in Trans. Linn. Soc. Lond. 27: 45, t. 17 (1869). —Ficalho, Pl. Uteis: 255 (1844). —Engl., Hochgebergsfl. Trop. Afr.: 336 (1892).—Knoblauch in Bot. Zentralbl. 60: 330 (1894). —Gilg in Engl., Pflanzenfam. 4, 2: 68 fig. 32 A–E (1895). —Hiern, Cat. Afr. Pl. Welw. 1: 710 (1898). —Gilg in Warb., Kunene-Samb.-Exped. Baum: 33 (1903). —Baker & N.E. Br. in F.T.A. 4, 1: 569 (1903). —Hill & Prain in F.C. 4, 1: 1118 (1909). —R.E. Fries, Wiss. Ergebn. Schwed. Rhod.-Kongo-Exped. 1911–1912, 1: 261 (1914). —Marais & Verdoorn in Fl. S. Afr. 26: 241 (1963). —Boutique in Fl. Afr. Centr., Gentianaceae: 23, t. 4 (1972). —Taylor in Garçia de Orta, Sér. Bot. 1: 79, t. 8, fig. 10–17 (1973). Type from Angola.
    Faroa boehmii Engl., Pflanzenw. Ost-Afr. C: 313 (1895). —Baker. & N.E. Br. in F.T.A. 4, 1: 568 (1903). Type from Tanzania.
    Faroa nyasica N.E. Br. in F.T.A. 4, 1: 568 (1903). Types: Malawi, W. of Lake Malawi, *Kirk* s.n. (K, syntype); *Livingstone* s.n. (K, syntype).

Perennial erect herb, 5–15 cm. tall, with many basal stems arising from a rootstock, rarely solitary. Stem simple or branched from the base, 4-ridged. Rosette leaves sometimes present at the time of anthesis, 15–90 × 3–10 mm., from ovate or oblong-ovate to oblong lanceolate or linear; stem leaves in 3–4 pairs, 4–15 × 1.5–3 mm., lanceolate to oblong-obovate. Inflorescence dense axillary and terminal clusters, simple or umbelliform, sometimes the terminal one head-like. Flowers blue, mauve to whitish; pedicels 0.3–0.7 mm. long, filiform. Calyx tube 1–1.25 mm. long, campanulate; lobes 1.5–2(3) × 0.7–0.9 mm., ovate to ovate-lanceolate, subacute or obtuse at the base, carinate at the apex, 1-nerved. Corolla tube 1.5–2.5 mm. long, cylindric; lobes 1.8–2 × 0.6–0.8 mm., elliptic-lanceolate, attenuate at the apex. Filaments 2–3 mm. long, filiform; anthers 0.6–0.8 mm. long, ellipsoid. Ovary 1.8–2 × 0.7–0.9 mm., cylindric-obovoid; style 3–6 mm. long, filiform; stigma subcapitate. Capsule 1.8–2.1 × 0.9–1 mm., obovoid-cylindric. Seeds c. 0.3 mm. in diam., subglobose, minutely scrobiculate.

Zambia. B: Kalabo, fl. & fr. 15.xi.1959, *Drummond & Cookson* 6478 (BR; COI; K; LISC). N: Kawambwa, fl. & fr. 15.xi.1957, *Fanshawe* 4045 (K). W: Mwinilunga, fl. & fr. 7.viii.1930, *Milne-Redhead* 856 (K). C: Lusaka Distr., Chongwe R., near Constantia (N. of Kasisi), fl. & fr. 11.x.1972, *Kornaś* 2374

(K). **Zimbabwe**. N: Between Wellesley and Darwendale, fl. & fr. 25.ix.1955, *Drummond* 4874 (K; LISC; PRE). C: Harare, fl. & fr. x.1919, *Eyles* 1842 (K). S: Masvingo, s.d. *Munro* 359A (BM), 580 (BM). **Malawi**. N: W. of Lake Malawi, fl. & fr. *Kirk* s.n. (K). C: Kasungu Nat. Park, fl. & fr. 8.ix.1972, *Pawek* 5703 (K). S: Zomba, fl. & fr. 1901, *Sharpe* 31 (K). **Mozambique**. N: Cuamba, fl. & fr. 20.ix.1935, *Torre* 685 (COI; LISC). Z: Between Pebane and Mocubela, fl. & fr. 25.x.1942, *Torre* 4691 (LISC). MS: Cheringoma, near Dondo, fl. & fr. 14.x.1944, *Mendonça* 2473 (LISC).

Also in Angola, Zaire, S. Africa and Tanzania. Damp or seasonally wet grassland, riverbanks and marshes; 450–1500 m.

4. **Faroa corniculata** Taylor in Garçia de Orta, Sér. Bot. **1**: 76, t. 7, fig. 7–10 (1973). TAB **14**. Type: Zambia, Sansia Falls, Kalambo R., *Richards* 24555 (K, holotype).

Succose, laxly caespitose perennial herb. Stem woody at the base, up to 10 cm. long, 4-ridged, almost winged. Leaves 10–35 × 2–7 mm., elliptic or oblong-elliptic, obtuse at the apex, slightly connate at the base. Flowers pale mauve to white-purplish arranged in axillary 2–4-flowered fascicles; pedicels up to 2 mm. long, filiform, 4-winged. Calyx tube 0.8–1 mm. long, campanulate-urceolate; lobes 2–2.5 × 0.8–1 mm., ovate, strongly winged on the keel, shortly cuspidate at the apex. Corolla tube 1.5–1.7 mm. long; lobes 1–1.2 × 0.3–0.5 mm., ovate, externally subapically corniculate. Filaments 0.25–0.5 mm. long, filiform; anthers 0.4–0.5 mm. long, cylindric-ovoid. Ovary 1.5–1.7 × 0.9–1 mm., obovoid-ellipsoid; style 0.9–1 mm. long; stigma shortly bilobed. Capsule 1.9–2 × 1–1.2 mm., ovoid. Seeds c. 0.3 mm. in diam., subglobose, minutely scrobiculate.

Zambia. N: Mbala Distr., Chilongowelo, fl. & fr. 8.iv.1952, *Richards* 1461 (K). Not known elsewhere. Rock crevices; c. 1500 m.

5. **Faroa axillaris** Baker in Kew Bull. **1898**: 158 (1898). —Baker & N.E. Br. in F.T.A. **4**, 1: 566 (1903). —Boutique in Fl. Afr. Centr., Gentianaceae: 24 (1972). —Taylor in Garçia de Orta, Sér. Bot. **1**: 74, t. 5 fig. 7–16 (1983). Type: Malawi, Misuku Plateau, *Whyte* s.n. (K, holotype).

Erect annual herb, 6–15 cm. tall. Stem more or less branched from the base, 4-ridged. Leaves 10–40 × 3–10 (12.5) mm., the lower obovate-oblong or oblong-spathulate, obtuse at the apex, attenuate at the base; the upper oblong-lanceolate, dilated, scarious, 5-nerved and connate at the base forming an involucre surrounding the inflorescence, secondary nerves with a coarse reticulation. Flower blue, to whitish, in axillary and terminal many-flowered fascicles; pedicels 0.5–1 mm. long, 4-ridged. Calyx tube 1–1.5 mm. long, campanulate; lobes 2.3 × 0.8–1.25 mm., obovate-oblong or oblong-subspathulate, obtuse at the apex, with 3-nerves from the base and anastomosing secondary nerves near the apex. Corolla tube 1.5–1.8(2) mm. long, papillose, with distinct scarcely elongated epidermis cells; lobes 1–1.5 × 1.1–1.4 mm., broadly ovate, subacute at the apex, externally coarsely papillose, internally smooth. Filaments 0.8–1 mm. long, filiform; anthers c. 0.5 mm. long, subglobose. Ovary 1.7–2 × 1–1.1 mm., obovoid; style 1.3–1.4 mm. long; stigma of 2 filiform recurved lobes. Capsule 2.3–2.5 × 1.6–1.8 mm., obovoid. Seeds c. 0.5 mm. in diam., subglobose, minutely scrobiculate.

Zambia. N: Kasama to Mpika Rd. near Chambeshi pontoon, fl. & fr. 29.iv.1962, *Richards* 16386 (K). W: Chati, fl. & fr. 26.v.1972, *Fanshawe* 11462 (K). **Malawi**. N: Chitipa Distr., Kaseye Mission, fl. & fr. 26.iv.1972, *Pawek* 5252 (K). **Mozambique**. N: Niassa, 8 km. E. of Mandimba, fl. & fr. 15.v.1961, *Leach & Rutherford Smith* 10851 (K; LISC).

Also in Zaire, Angola and Tanzania. Open deciduous woodland in dry sandy or gravelly soil; 700–2100 m.

6. **Faroa involucrata** (Klotzsch) Knoblauch in Bot. Zentralbl. **60**: (1894). —Engl. Pflanzenw. Ost- Afr. **C**: 313 (1895). —Baker & N. E. Br. in F.T.A. **4**, 1: 566 (1903). —Taylor in Garçia de Orta, Sér. Bot. **1**: 73, t. 5 fig. 1–6 (1973). Type: Mozambique, Boror, *Peters* s.n. (B, holotype†; K, fragment).

Sebaea involucrata Klotzsch in Peters, Reise Mossamb., Bot.: 271 (1861). —Welw. in Trans. Linn. Soc. **27**: 47 (1869). Type as above.

Erect, annual herb, up to 30 cm. tall. Stem more or less branched from the base, 4-ridged. Leaves 10–30 × 3–9 mm., the lower obovate-oblong or oblong-spathulate, obtuse at the apex, attenuate at the base; the upper oblong-lanceolate, acute at the apex, dilated, scarious, very broadly 7-nerved and connate at the base, forming an involucre surrounding the inflorescence, the secondary nerves finely reticulated. Flowers blue, in axillary and terminal many-flowered fascicles; pedicels 0.5–1 mm. long, 4-ridged. Calyx tube 2–2.5 mm. long, campanulate; lobes 1.7–2 × 1.3 mm., oblong-ovate or spathulate-

44

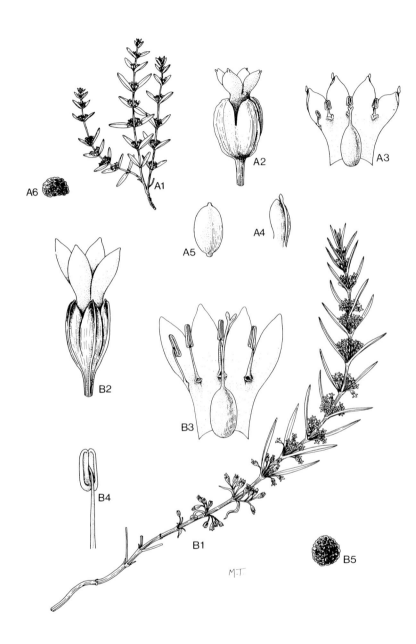

Tab. 14. A.—FAROA CORNICULATA. A1, fertile branch (×$\frac{1}{2}$); A2, flower (×6); A3, corolla opened
out to show pistil and stamens (×6); A4, dorsal view of corolla lobe (×6); A5, capsule (×6); A6,
seed (×12), A1–6 from *Richards* 15127. B.—FAROA ALATA. B1, part of stem (×$\frac{1}{2}$); B2, flower
(×6); B3, corolla opened out to show pistil and stamens (×6); B4, stamen (×6); B5, seed (×12),
B1–5 from *Phipps & Vesey-FitzGerald* 3192.

ovate, obtuse, mucronate, with recurved spreading or squarrous apices, carinate down the back, with a conspicuous midrib and numerous sub-parallel nerves from the base and anastomosing secondary nerves near the apex. Corolla tube 2.5–3 mm. long, smooth with obscure elongated epiderm cells; lobes 1.7–2(2.5) × 1.3–1.5 mm., broadly ovate, subacute at the apex, externally smooth, and internally minutely papillose. Filaments 1.6–2 mm. long, filiform; anthers 0.75 mm., ellipsoid. Ovary 2.8–3 × 1.6–1.8 mm., obovoid-ellipsoid; style 1.5–1.75 mm. long; stigma of 2 filiform recurved lobes. Capsule 3.5 × 1.9–2.1 mm., obovoid. Seeds 0.6–0.8 mm. in diam., subglobose, minutely scrobiculate.

Mozambique. N: 10 km. N. of Nampula, fl. & fr. 13.vi.1944, *Gomes e Sousa* 3332 (PRE). Z: Boror, *Peters* s.n. (K, fragment); Ile s.d., *Torre* 5574 (K; LISC).
Not known elsewhere. Rocky situations, apparently associated with mineral outcrops; at low altitudes.

7. **Faroa amara** Gilg ex Baker & N.E. Br. in F.T.A. **4**, 1: 567 (1903). —R.E. Fries, Wiss. Ergebn. Schwed. Rhod.-Kongo-Exped. 1911–1912, **1**: 261 (1914). —Taylor in Garçia de Orta, Sér. Bot. **1**: 78, t. 9 fig. 6–11 (1973). Type from Angola.

Caespitose annual herb. Stem up to 6–7(12) cm. long, branched, 4-ridged, almost winged. Leaves subsessile, sometimes scarcely exceeding the subtended flower fascicle, 10–35 × 3–10 mm., spathulate-ovate or oblanceolate, cuneate at the base. Flowers mauve, purplish to blue, in axillary clusters, which are continuous and form a dense oblong terminal inflorescence; pedicels 2.5–4 mm. long; filiform. Calyx tube 0.8–1 mm. long, campanulate; lobes 1.5–2 × 0.7–1.2 mm., elliptic or oblong-lanceolate, scarcely winged on the keel at the apex, obtuse or subapiculate, 1-nerved. Corolla tube 1.5–2 mm. long, cylindric, smooth; lobes 0.9–1.2 × 0.5–0.6 mm., oblong-ovate, acute. Filaments 0.25 mm. long; anthers 0.25–0.3 mm. long, ovoid-ellipsoid. Ovary 1.8–2 × 0.8–1 mm., obovoid-cylindric; style 0.8–1.5 mm. long; stigma sub-bilobed. Capsule 2–2.2 × 0.9–1.2 mm., obovoid. Seeds c. 0.25 mm. in diam., globose, minutely scrobiculate.

Zambia. N: Shiwa Ngandu, fl. & fr. 4.vi.1958, *Robinson* 1588 (K). W: Mufulira, fl. & fr. 31.v.1934, *Eyles* 8246 (K). **Zimbabwe**. N: Gokwe, Sengwa Research St., fl. & fr. 20.iv.1976, *Guy* 2429 (PRE). W: Matobo, fl. & fr. iv.1955, *Miller* 2790 (K). C: Harare, Twentydales, fl. & fr. 28.iv.1948, *Wild* 2522 (K). S: c. 32 km. N. of Masvingo, fl. & fr. 4.v.1962, *Drummond* 7954 (K).
Also in Tanzania and Angola. Moist sandy soil in open rocky situations; 1000–1900 m.

8. **Faroa minutiflora** Taylor in Garçia de Orta, Sér. Bot. **1**: 79 t. 9 fig. 12–14 (1973). Type: Zambia, Kasama Distr., Mungwi to Kasama, *Richards* 16426 (K, holotype).

Erect annual herb, 4–8 cm. tall. Stem unbranched, 4-ridged. Leaves 5–15 × 1–1.5 mm., linear to linear-lanceolate, acute at the apex, connate at the base, thinly fleshy. Flowers cream, in axillary 3–10-flowered fascicles; pedicels 1.5 mm. long. Calyx campanulate, lobes 0.8–1 × 0.8 mm. ovate, carinate at the apex, 1-nerved. Corolla tube 0.4–0.6 mm. long, smooth; lobes 0.4–0.6 × 0.35 mm., ovate-deltate. Filaments 0.3–0.4 mm. long, filiform; anthers 0.2–0.25 mm., ovoid-subglobose. Ovary 1.7 × 0.8 mm., obovoid; style c. 0.3 mm. long, filiform; stigma subcapitate. Capsule 1.9–2 × 0.8–1 mm., obovoid. Seeds c. 0.3 mm. in diam., depressed globose, minutely scrobiculate.

Zambia. N: Kasama Distr., Mungwi to Kasama, fl. & fr. 30.iv.1962, *Richards* 16426 (K, holotype).
Not known elsewhere. Damp sand, among rocks; 1260 m.

9. **Faroa pusilla** Baker in Kew Bull. **1894**: 26 (1894). —Baker & N.E. Br. in F.T.A. **4**, 1: 568 (1903). —Hutch. & Dalz., F.W.T.A. **2**, 1: 183 (1931). —Taylor in F.W.T.A., ed. 2, **2**: 302 (1963). —Boutique in Fl. Afr. Centr., Gentianaceae: 26 (1972). —Taylor in Garçia de Orta, Sér. Bot. **1**: 78, t. 9 fig. 1–5 (1973). Type from Nigeria.
Faroa schweinfurthii Engl. & Knoblauch in Bot. Zentralbl. **60**: 330 (1894). Type from Sudan.

Erect annual herb, 2–10 cm. tall. Stem not or sparsely branched, 4-angled, widely winged. Leaves 5–35 × 1.5–15 mm., oblong-elliptic or obovate-oblong, lanceolate to linear-lanceolate, acute or acuminate at the apex, cuneate at the base. Flowers blue-violet to whitish, in axillary and terminal more or less 10-flowered fascicles; pedicels 1.5–2 mm. long, filiform. Calyx tube 1–1.25 mm. long, campanulate; lobes 1–1.5 × 0.8–1.2 mm., ovate, acute, prominently carinate at the apex, with a short erect apiculus, 1-nerved. Corolla tube 1.6–1.8 mm. long, cylindric, papillose; lobes 1–1.25 × 0.5–0.7 mm., obovate, subacute. Filaments c. 0.5 mm. long, filiform; anthers c. 0.25 mm. long, ellipsoid. Ovary 1.7–2 ×

0.75–1 mm., obovoid; style 0.6–0.9 mm. long; stigma of 2 filiform lobes, recurved. Capsule 1.8–2 × 1–1.25 mm., obovoid. Seeds c. 0.3 mm. in diam., subglobose, minutely scrobiculate.

Zambia. B: Mongu, fl. & fr. 11.iv.1966, *Robinson* 6926 (K). W: 11 km. N. of Chingola, fl. & fr. 4.v.1960, *Robinson* 3704 (K). C: Serenje Distr., Kundalila Falls, fl. & fr. 5.v.1972, *Kornaś* 1665 (K).
Widespread throughout tropical Africa. Damp or drying open woodland or grassland; from near sea level to 1000 m.

10. **Faroa alata** Taylor in Garçia de Orta, Sér. Bot. **1**: 75, tab. 7 fig. 1–3 (1973). TAB. **14**. Type: Zambia, Mporokoso, Mweru-Wantipa, near Muzombwe, *Phipps & Vesey-FitzGerald* 3192 (K, holotype; LISC; PRE).

Succose, branched, perennial herb. Stem woody at the base, up to 35 cm. long, 4-ridged, almost winged. Leaves 5–35 × 1.5 mm., oblong-elliptic, obovate-oblong to linear-lanceolate, acute at the apex, cuneate at the base. Flowers white, in 6 flowered fascicles; pedicels 3–6 mm. long, filiform, 4-ridged. Calyx tube 2–2.3 mm. long, campanulate; lobes 2.2–2.5 × 0.9–1 mm., ovate, apiculate, winged on the keel, 1-nerved. Corolla tube 3–4.2 mm. long; lobes 2.5–3 × 1.25–1.5 mm., ovate lanceolate, subacute. Filaments 1.7–2 mm. long, filiform; anthers 1–1.2 mm. long, ellipsoid. Ovary 2.3–2.5 × 1–1.3 mm., obovoid-cylindric; style 4.2–4.5 mm. long; stigma shortly bilobed. Capsule 2.5–2.75 × 1–1.5 mm., ovoid-ellipsoid. Seeds c. 0.5 mm. in diam., globose, minutely scrobiculate.

Zambia. N: Mporokoso, Mweru-Wantipa, near Muzombowe, *Phipps & Vesey-FitzGerald* 3192 (K; LISC; PRE).
Not known elsewhere. Rocky pavement; c. 1000 m.

11. **Faroa hutchinsonii** Taylor in Garçia de Orta, Sér. Bot. **1**: 76, t. 7 fig. 4–6 (1973). Type: Zambia, Serenje Corner, *Hutchinson & Gillett* 3709 (BM; K, holotype).
    Faroa amara sensu Hutch., Botanist in South Afr.: 501 (1946) non Gilg ex Baker. & N.E. Br. (1903).

Succose caespitose perennial herb. Stem up to 7 (10) cm. long, branched, 4-ridged, almost winged. Leaves 6–10 × 2–4 mm., oblong-ovate or oblong-lanceolate, cuneate at the base, obtuse at the apex, fleshy, subsessile, sometimes scarcely exceeding the subtended fascicle of flowers. Flowers blue, in axillary 3–4-flowered fascicles which are congested to form a subcapitate inflorescence; pedicels 2–4 mm. long, filiform. Calyx tube 0.8–1 mm. long, campanulate; lobes 1.5–2 × 0.8–1 mm., ovate or ovate-oblong, distinctly winged or the upper part of the keel (at least the outward pair), apiculate, 1-nerved. Corolla tube c. 2.5 mm. long, cylindric; lobes 2–2.5 × 0.9–1 mm., ovate-oblong, subacute. Filaments 1.9–2.2 mm. long; anthers c. 0.5 mm. long, ovoid. Ovary 2–2.4 × 1–1.25 mm., obovoid-cylindric; style 2–2.5 mm. long; stigma obsolete. Capsule 2–2.5 mm. long, obovoid. Seeds c. 0.25 mm. in diam., globose, minutely scrobiculate.

Zambia. C: Serenje Corner, fl. & fr. 16.vii.1930, *Hutchinson & Gillett* 3709 (BM; K).
Not known elsewhere. Sandstone rocks; 2160 m.

12. **Faroa fanshawei** Taylor in Bull. Bot. Nat. Belg. **41**: 267 (1971). —Boutique in Fl. Afr. Centr., Gentianaceae: 27 (1972). —Taylor in Garçia de Orta, Sér. Bot. **1**: 77, t. 7 fig. 11–17 (1973). Type: Zambia, Mupata Gorge, *Fanshawe* 31 (K, holotype).

Annual herb. Stem decumbent or erect up to 18 cm. long, more or less branched from above the base, 4-ridged. Leaves 10–30 × 3–10 mm., oblong-elliptic or lanceolate, obtuse at the apex, cuneate or sometimes slightly cuneate at the base, thinly succose. Flowers white, in axillary 3–9-flowered usually lax (except the lowest ones) fascicles; pedicels up to 8 mm. long, filiform, 4-ridged, not papillate. Calyx tube 0.6–0.9 mm. long, campanulate; lobes 1.3–1.7 × 0.9–1.2 mm., ovate, winged on the upper part of the keel, 1-nerved. Corolla tube cylindric 2–2.5 mm. long; lobes 2–2.5 × 0.8–1 mm., oblong-ovate, obtuse at the apex. Filaments 2.2–2.8 mm. long, filiform; anthers 0.6–0.8 mm. long, ellipsoid. Ovary 1.7–2 × 0.8–1 mm., obovoid; style 3–4 mm. long, filiform; stigma subcapitate. Capsule 1.8–2.2 × 1 mm., obovoid. Seeds c. 0.5 mm. in diam., globose, minutely scrobiculate.

Zambia. W: Mupata Gorge (W. of Luanshya), fl. & fr. 25.v.1953, *Fanshawe* 31 (K).
Also in Zaire. Quartzite kopjes.

13. **Faroa affinis** De Wild. in Ann. Mus. Congo Bot., Sér. 4, **1**: 99, t. 11, fig. 14–21 (1903). —Baker & N.E. Br. in F.T.A. **4**, 1: 567 (1903). —Boutique in Fl. Afr. Centr., Gentianaceae: 28 (1972). —Taylor in Garçia de Orta, Sér. Bot. **1**: 77, t. 8 fig. 1–4 (1973). Type from Zaire.
  Faroa wellmanii Prain in Kew Bull. **1908**: 260 (1908). Type from Angola.

Erect annual herb, 5–20 cm. tall. Stem not or more or less branched from above the base, 4-ridged. Leaves 3–25 × 1.5–8 mm., lanceolate or oblanceolate, subacute at the apex, cuneate at the base, thinly fleshy. Flowers mauve, blue to purple, in axillary many-flowered fascicles; pedicels 2–5 mm. long, papillate. Calyx tube 0.5–0.7 mm. long, campanulate; lobes 1.5–2.5 × 0.5–0.9 mm., lanceolate to oblong-lanceolate, not winged, with more or less denticulate margins, 1-nerved. Corolla tube 1.2–1.5 mm. long, cylindric, papillose; lobes 2–2.7 × 0.6–0.75 mm., oblong-lanceolate, obtuse. Filaments 4–4.5 mm. long, filiform; anthers c. 0.5 mm. long, ovoid. Ovary 1.5–1.8 × 0.6–0.7 mm., cylindric-obovoid; style 4–4.3 mm. long, filiform; stigma subcapitate. Capsule 1.8–2 × 0.7–0.8 mm., cylindric-obovoid. Seeds c. 0.3 mm. in diam., subglobose, minutely scrobiculate.

Zambia. N: Kawambwa, fl. & fr. 23.viii.1957, *Fanshawe* 3573 (K). W: Mufulira, fl. & fr. 17.v.1934, *Eyles* 8297 (K). E: Lundazi, fl. & fr. 22.vi.1967, *Hilundu* 31 (K; LISC).
Also in Angola and Zaire. Damp grasslands; 1200–1500 m.

10. PYCHNOSPHAERA Gilg

Pychnosphaera Gilg in Warb., Kunene-Samb.-Exped. Baum: 333, t. 4 (1903).

Annual or perennial herbs. Stems erect, unbranched or corymbosely branching at the apex. Basal leaves oblong-lanceolate or oblong-elliptic; stem leaves from elliptic (lower), oblong-lanceolate (the middle ones), to linear or linear-oblong (upper), entire. Inflorescence in dense bracteate heads, solitary or corymbosely arranged at the end of the stems and branches. Flowers trimerous. Calyx unequally tripartite; lobes carinate. Corolla tube cylindric, with 3 spreading lobes. Stamens 3; filaments filiform, inserted at the sinuses of the corolla lobes, dilated into a small hood at their base; anthers cylindric-ovoid. Ovary narrowly ellipsoid, unilocular; ovules numerous; style filiform; stigmas 2, linear-oblong. Capsule oblong-elliptic, bivalved.

A monotypic tropical African genus.

Pychnosphaera buchananii (Baker) N.E. Br. in F.T.A. **4**, 1: 565 (1903). —Taylor in F.W.T.A. ed. 2, **2**: 302 (1963). —Boutique in Fl. Afr. Centr., Gentianaceae: 3 t. 1 (1972). TAB. **15**. Type: Malawi, *Buchanan* 1135 (BM; K, holotype).
  Faroa buchananii Baker in Kew Bull. **1894**: 26 (1894). Type as above.
  Pychnosphaera quarrei De Wild. in Contr. Fl. Nat. Suppl. **4**: 82 (1931). Type from Zaire.
  Pychnospharea vanderystii De Wild. in Pl. Bequaert. **2**: 107 (1923). Type from Zaire.
  Pychnospharea trimera Gilg in Warb., Kunene-Samb.-Exped. Baum: 333, t. 4 (1903). Type from Angola.

Annual or perennial herb up to 1.5 m. tall. Stem glabrous, unbranched or corymbosely branches at the apex, longitudinally 2-ridged. Basal leaves sometimes crowded, 1.5–5.5 × 0.5–0.8 cm., oblong-lanceolate or oblong-elliptic, stem-leaves opposite, 3–5.5 × 0.2–0.3 (0.5) cm., from elliptic to linear or linear-oblong, usually attenuate towards the apex or obtuse; base decurrent as ridges. Inflorescence in dense bracteate heads (bracts 3–6 × 2.5–3 mm., elliptic, sub-ovate, apiculate) solitary or corymbosely arranged to the apex of the stem and branches. Flowers from pink, mauve and blue-violet to white, subsessile; pedicels 0.3–1.2 mm. long. Calyx tripartite; lateral sepals 3–3.5 × 1.2 mm., oblong-elliptic, acute carinate; dorsal sepal 1.3–1.5 × 0.7 mm., oblong-obovate, marginate or obscurely 3-lobulate at the apex. Corolla tube 3–3.5 mm. long; lobes 2.5 × 1.5 mm., elliptic or ovate lanceolate, acute. Filaments 2.7–3 mm. long, linear; anthers 0.35–0.7 mm. long, cylindric-ovoid, minutely apiculate. Ovary 1–1.5 × 0.25 mm., cylindric-ellipsoid; style 2.5–2.7 mm. long, filiform; stigma of 2 linear-oblong branches, 1.3 mm. long. Capsule 2–2.5 × 1 mm., cylindric-ellipsoid. Seeds 9.25–0.3 mm. in diam., ovoid-ellipsoid to subglobose, brown; testa faveolate.

Botswana. N: Okavango Swamps, island on Moandchira R., fl. & fr. 8.ii.1972, *Biegel & Russell* 3902 (K; LISC; SRGH). **Zambia**. N: Mbala Distr., bog source of Inono stream, fl. & fr. 14.vi.1952, *Richards* 1432 (K). W: Mwinilunga, fl. & fr. 8.vi.1963, *Loveridge* 861 (K; LISC; PRE; SRGH). C: Chakwenga Headwaters, E. of Lusaka, fl. & fr. 27.iii.1965, *Robinson* 6478 (K). **Malawi**. C: Nkhota Kota, fl. & fr. 16.ii.1944, *Benson* 571 (K).
Tropical Africa from Guinea and Angola to Tanzania. In swamp forest marsh and river banks.

48

Tab. 15. PYCHNOSPHAERA BUCHANANII. 1, habit (× ½); 2, dorsal sepal (× 6); 3, flower, note lateral sepals (× 6); 4, corolla opened out to show pistil and stamens (× 6); 5, seed (× 25), all from *Richards* 1432.

11. ENICOSTEMA Blume

Enicostema Blume in Bijd. Fl. Nederl. Indie **14**: 848 (1826). —Raynal in Adansonia, Sér. 2, **9**: 75 (1969).

Annual or perennial herbs rarely suffrutescent. Stems erect, simple or branched at the base. Leaves sessile or narrowed into a petiole-like base, oblong-lanceolate to linear, usually the lower ones larger, entire 3-nerved. Flowers (3)4–5(6)-merous, in axillary clusters. Calyx with 3 short basal glands, irregular, rarely subregular by reduction of two lobes; tube campanulate; lobes ovate-lanceolate, obovate to subcircular or triangular-lanceolate. Corolla regular, tubuliform; lobes patent at anthesis, shorter than the tube. Stamens inserted in the corolla tube; filaments with a double-hooded scale at the base; anthers erect, not twisted, acute. Ovary narrowly ellipsoid, unilocular; ovules numerous; style shorter than the ovary, subulate; stigma subcapitate. Capsule obovoid, septicidally, bivalved. Seeds numerous, subglobose, faveolate.

A pantropical genus of three species, one from America, one from Madagascar and one from Africa and Asia.

Enicostema axillare (Lam.) Raynal in Adansonia, Sér. 2, **9**: 75, t. 2 fig. 1–2 (1969). TAB. **16**. Type from India.

Gentiana axillaris Lam., Ill. Gen. 1, **2**: 487 (1793) non Rafinesque 1828, nec (F.W. Schmidt; 1794) Reichenb. (1828). Type as above.

Enicostema verticillatum (L.) Engl. ex. Gilg in Engl. & Prantl, Pflanzenfam. 4, **2**: 67 (1895) pro parte.

Enicostema littorale sensu Baker & N. E. Br. in F.T.A. **4**, 1: 563 (1903). —Hill & Prain in F.C. **4**, 1: 1117 (1909). —Hutch. & Dalz. in F.W.T.A. **2**, 1: 183 (1931) non Blume (1826).

Exacum hyssopifolium Willd. in Sp. Pl. **1**: 640 (1798). Lectotype from India.

Hippion verticillatum sensu Hiern., Cat. Afr. Pl. Welw. **1**, 3: 711 (1898) non (L.) Schmidt (1796).

Enicostema hyssopifolium (Willd.) Verdoorn in Bothalia **7**: 142 (1961). —Taylor in F.W.T.A. ed. 2, **2**: 302 (1963). —Marais & Verdoorn in Fl. S. Afr. **26**: 238 fig. 35, 1 (1963). —Friedrich-Holzhammer in Merxm., Prodr. Fl. SW. Afr. **110**, Gentianaceae: 2 (1967). —Ross in Bot. Surv. Mem. **39**: 279 (1972). Lectotype as above.

Erect perennial herb, 5–30 cm. tall, simple or branched at the base. Stem cylindric, glabrous with a decurrent ridge below each leaf. Leaves sessile (sometimes narrowed into a petiole-like base), longer than the internodes; lamina (1)5–8 × 0.3–1 cm., linear to lanceolate or narrowly oblong, entire, obtuse and mucronate at the apex, somewhat narrowing towards the base, 3-nerved from the base, glabrous. Inflorescence in many flowered axillary clusters, numerous in the axils of each pair of leaves. Flowers white with green lines, drying yellowish, sessile or subsessile; bracts long, shorter than the calyx, lanceolate-acuminate, carinate. Calyx tube 1–2 mm. long; lobes usually unequal, 0.7–1.5(2) × 0.4–0.7 mm., triangular to lanceolate, acute at the apex and narrowly scarious at the margin, or obovate to subcircular, obtuse and mucronate at the apex, with wide scarious margin. Corolla tube 3.5–6 mm. long; lobes 1.5–2 × 0.7–1 mm., ovate and abruptly narrowing to an acute or mucronate apex. Stamens inserted below the sinuses, just above the middle of the tube; filaments 1.5–2.3 mm. long, with a double hood at the insertion point; anthers 1 mm. long, erect, shortly apiculate. Ovary 5–6 × 1 mm., ovoid; style 2–2.5 mm. long, subulate; stigma subcapitate. Capsule 3–4.5 × 2–2.5 mm., obovoid. Seeds 0.4–0.5 mm. in diam., subglobose, reticulate faveolate.

Widespread throughout the tropics of the world.

Calyx lobes triangular or lanceolate, acute at the apex and narrowly scarious at the margin - - - - - - - - - - - - - - - subsp. *axillare*
Calyx lobes obovate to subcircular, obtuse and mucronate at the apex, with a wide scarious margin - - - - - - - - - - - subsp. *latilobum*

Subsp. **axillare**

Leaf lamina obtuse and mucronate at the apex. Bracts lanceolate acuminate, acute at the the apex, narrowly membranous at the margin. Calyx lobes triangular or lanceolate, acute at the apex and narrowly scarious at the margin. Capsule 3–4.5 mm. long.

Botswana. N: Khardoum Valley, 30 km. E. of Namibia border, fl. & fr. 15.iii.1965, *Wild & Drummond* 7055 (K; SRGH). **Zambia**. S: Livingstone, Zambezi R., fl. & fr. 17.i.1929, *Blenkiron* s.n.

50

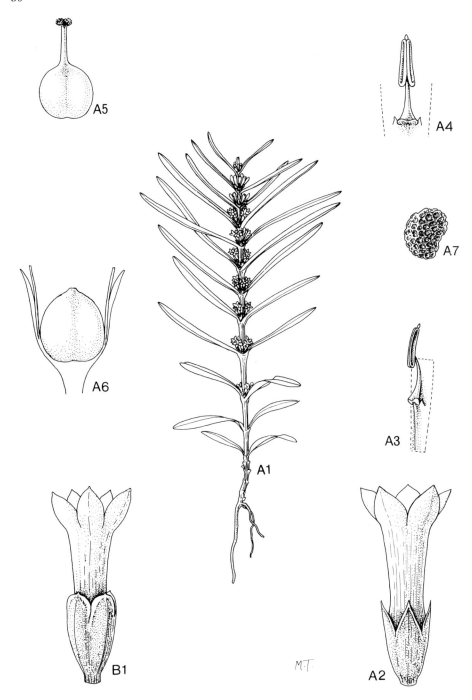

Tab. 16. A. —ENICOSTEMA AXILLARE subsp. AXILLARE. A1, habit (× ½); A2, flower (× 5); A3, stamen, lateral view (× 10); A4, stamen frontal view (× 10); A5, pistil (× 10); A6, capsule (× 10); A7, seed (× 25), all from *Wild & Drummond* 7055. B. —ENICOSTEMA AXILLARE subsp. LATILOBUM. B1, flower (× 5), from *Kirk* s.n.

(PRE). **Zimbabwe**. N: Mt. Darwin, Mzarabani T.T.L., Musingwa R., fl. i.v.1972, *Mavi* 1362 (K; LISC; SRGH). W: Hwange Game Reserve, Main Camp, fl. 20.ii.1956, *Wild* 4740 (K; PRE; SRGH). E: Chipinge, fl. 13.i.1957, *Phipps* 124 (K). S: Mwenezi, near Yangambi R., c. 8 km. W. of Mateke Hills, fl. 2.v.1958, *Drummond* 5531 (K; SRGH). **Malawi**. S: Nsanje R., Shin, Old Nature Garden fl. i.iv.1933, *Lawrence* 41 (K). **Mozambique**. N: Pemba, between Ancuabe and Metuge, fl. & fr. 7.ix.1948, *Barbosa* 2801 (LISU). Z: Namagoa, Mocuba, iii.1943, *Faulkner* 162 (K; COI; LISC; SRGH). T: Baroma, Zambezi R., fl. 9.vii.1950, *Chase* 2666 (K; LISC; SRGH). MS: Beira, Gorongosa Nat. Park, Rd. 5 Area, fl. ii.1969, *Tinley* 1691 (K; LISC). GI: Guija Prov., Limpopo R., fl. & fr. vii.1915, *Gazaland Exp.* 15804 (PRE). M: Maputo, Costa do Sol, fl. 7.v.1971, *Marques* 2267 (BM; COI; K; LISC; LMU).

Tropical Asia and Africa. Grasslands, in turf hollows, sandy sloped, side roads and along rivers.

Subsp. **latilobum** (N.E. Br.) Raynal in Adansonia Sér. 2, **9**: 77 t. 2 fig. 3; t. 3 fig. A (1969). TAB. **16**. Type from Kenya.

  Enicostema latiloba N.E. Br. in F.T.A. **4**, 1: 564 (1903). Type as above.
  Enicostema verticillatum (L.) Engl. ex Gilg in Engl. & Prantl, Pflanzenfam. 4, **2**: 67 pro parte; 68, fig. 31 (1895) pro parte quoad B–E.

Leaf lamina acute or obtuse and mucronate at the apex. Bracts broadly obovate or subobovoid-obovate, acute or obtuse and apiculate at the apex, with broad membranous margin. Calyx lobes obovate to subcircular, obtuse and mucronate at the apex, with a wide scarious margin. Capsule 3–4 mm. long.

  Mozambique. N: Rovuma Bay, fl. iii.1861, *Kirk* s.n. (K).
  Also in Kenya and Tanzania Coast. In Grossypium plantation and along water courses.

E. axillare subsp. *littorale* (Blume) Raynal does not occur in Africa. It is a Malaysian taxon (Java, Madura, Lombok, Sumbawa, Sumba and Timor) which occupies different habitats from the African ones (swamp, margins of streams, along roadsides, often not far from the seaside, whereas the African the subspecies are plants adapted to much more varied habitats) and has leaves acute-acuminate at the apex and capsule 4–6 mm. long, whereas the African subspecies has leaves obtuse and mucronate at the apex and the capsule 3–4 mm. long.

114. MENYANTHACEAE

By Barbara Mackinder

Aquatic herbs, with tufted rootstock or horizontal creeping rhizomes. Leaves floating or submersed, alternate or in a rosette, simple or compound, entire or crenulate; stipules small, rounded, scarious, or inconspicuous. Flowers in panicles, racemes, fascicles or solitary; actinomorphic, generally 5-merous, hermaphrodite or unisexual. Sepals free or almost so, imbricate in bud. Corolla gamopetalus, often with a short tube, lobes membranous, often fimbricate or pilose. Stamens as many as corolla lobes and alternate with them; anthers dithecous, opening by longitudinal slits. Ovary superior, unilocular with 2(5) parietal placentas. Fruit a capsule, dehiscent or not. Seeds often numerous, with endosperm.

A family of 5 genera, with 40–50 species worldwide. Sometimes included in the *Gentianaceae*, but Cronquist, in "An intergrated system of classification of Flowering Plants: 900–903 (1981)", lists a number of characters in the anatomy, embryology, chemistry and palynology, which suggest the *Menyanthaceae* are a distinctive family, possibly better referred to the *Solanales* than the *Gentianales*.

NYMPHOIDES Seguier

Nymphoides Seguier, Pl. Veron. **3**: 121 (1754). —A. Raynal in Mitt. Bot. Staatss. München **10**: 122–134 (1971); in Adansonia, Sér. 2, **14**: 227–270, 405–458 (1974); in Fl. Afr. Centr., Menyanthaceae: 1–16 (1975).

  Limnanthemum S. Gmelin in Nov. Comm. Acad. Petrop. **14**, 1: 527 (1769).

Aquatic herbs, annual or perennial. Leaves simple, circular to ovate, with a sinus at the base, in a rosette on the stock, or alternate along floating stolons. Flowers clustered at the nodes, pedicellate, subtended by a small bract, 5(8)-merous, sometimes heterostylous, elsewhere sometimes unisexual. Capsule developing under water, splitting into valves from base or apex. Seeds round to oval, flattened or convex, smooth to tuberculate.

Worldwide, with 9 species in tropical Africa. This account is largely based on the meticulous studies of the genus by Dr. Aline Raynal cited above. The flowers of the genus are very delicate, so that a number of distinctive features elucidated by Dr. Raynal had been overlooked previously. Collectors are encouraged to make careful notes of flower colour and hair-arrangement, and also to preserve flowers in spirit. More than one species may be found growing together.

1. Leaf lamina, mostly ovate, 1.2–1.7 times as long as broad, 2–5.8 × 1.5–4.7 cm., with a short gaping sinus 0.8–1.2 cm. long, membranous petiole 1–3 mm. long, slender annual without stolons   -   -   -   -   -   -   -   -   -   -   -   -   -   -   -   -   2
– Leaf lamina mostly subcircular or somewhat broader than long, with a deep narrow closed sinus, ± leathery; petiole 3–55 mm. long; perennial with or without stolons or annual with stolons linking floating leaves   -   -   -   -   -   -   -   -   -   -   -   -   -   3
2. Pedicels c. 20 mm. long; corolla yellow; seed subcircular in outline with strongly convex faces   -   -   -   -   -   -   -   -   -   -   -   -   -   -   6. milnei
– Pedicels 7–10(15) mm. long; corolla white; seeds ovate in outline, with flat faces   -   -   -   -   -   -   -   -   -   -   -   -   -   7. tenuissima
3. Corolla white, lobes entirely covered by hairs inside, tube c. 7–9 mm. long yellow; calyx 5–9 mm. long   -   -   -   -   -   -   -   -   -   -   -   -   -   4
– Corolla yellow, lobes fimbriate-laciniate along margins and midvein inside (hairy, sometimes more so at throat), tube 1.5–6 mm.; calyx 3–6.5 mm. long   -   -   -   -   -   5
4. Corolla lobes ± as long as tube; sepals appressed to corolla tube; capsule subglobose, ± as long or longer than calyx; carpels (2)3(5)   -   -   -   -   -   -   1. brevipedicellata
– Corolla lobes ± twice as long as tube; sepals spreading at anthesis (later closing around capsule); capsule ovoid shorter than calyx   -   -   -   -   -   -   -   -   -   2. indica
5. Corolla less than 10 mm. long, homostylous (stigma at same level as anthers); pedicels 5–25(30) mm. long; capsule globose   -   -   -   -   -   -   -   -   -   -   5. rautanenii
– Corolla ± 10–15 mm. long; heterostylous (stigma well above or below anthers); pedicels mostly 25–70 mm. long   -   -   -   -   -   -   -   -   -   -   -   -   -   6
6. Corolla pale yellow, with a tube 5–6 mm. long; few stolons produced; seeds smooth to ± echinulate   -   -   -   -   -   -   -   -   -   -   -   - 3. thunbergiana
– Corolla golden-yellow with a tube 3–3.5 mm. long; stolons freely produced in mature plants; seeds smooth to echinulate, when well developed often tuberculate -   -   -   4. forbesiana

1. **Nymphoides brevipedicellata** (Vatke) A. Raynal in Mitt. Bot. Staatss. München **10**: 125, map 2 (1971); in Adansonia, Sér. 2, **14**: 414, t. 17 (1974); in Fl. Afr. Centr., Menyanthaceae: 14, t. 6 (1975). Type from Ethiopia.
     Limnanthemum brevipedicellatum Vatke in Linnaea **40**: 220 (1876). Type as above.
     Limnanthemum abyssinicum N.E. Br. in F.T.A. **4**, 1: 584 (1904). Type from Ethiopia.
     Nymphoides indica sensu Marais & Verdoorn in Fl. S. Afr. **26**: 243 (1963) pro parte. — Friedrich-Holzhammer in Merxm., Prod. Fl. SW. Afr. **111**: 1 (1967) pro parte.

Perennial; stock short thick, floating stolons numerous, 3–5 mm. thick in life. Petiole 1.2–1.5(4) cm.; lamina 6–10(23) cm. in diam., subcircular or the larger ones somewhat ovate-circular. Flowers 10–25 homostylous; pedicels 2–6 cm. long. Calyx 5–6 mm. long, lobes appressed to corolla tube. Corolla tube yellow 7–9 mm. long; lobes white, densely and uniformly hairy inside ± as long as the tube. Carpels and stigmas (2)3(5). Capsule 5–8 mm. diam., subglobose ± as long as the calyx, 15–80 seeded. Seeds 1.6–1.95 × 1.4–1.7 mm., lenticular ± tuberculate.

Botswana. N: Ngamiland Distr., Okavango R. 1 km. S. of Shakawe, fl. & fr. 24.iv.1975, *Gibbs-Russell* 2779 (K; SRGH). **Zambia.** B: 8 km. N. of Senanga, fl. & fr. 30.vii.1952, *Codd* 7254 (K; PRE). N: Mbala Distr., Kawimbe, Lumi R., fl. & fr. 30.v.1961, *Richards* 15169 (K). W: Kitwe, fl. & fr. 29.viii.1955, *Fanshawe* 2439 (K; NDO). C/S: Kafue Nat. Park, Kafue R., cult. Kew, fl. viii.1966, *Mitchell* H 3380/66 (K). S: 19 km. N. of Choma, fl. & fr. 24.iii.1957, *Robinson* 2190 (K; SRGH). **Zimbabwe.** W: Shangani Reserve, fr. 5.iv.1951, *West* 3156 (K; SRGH). C: Marondera Distr. Chikokorona Pan, fl. & fr. 29.iv.1982, *Gibbs-Russell* 1990 (SRGH; PRE). **Malawi.** C: Kasungu Game Res., Lingadzi Dambo, fl. & fr. 22.vi.1970, *Brummitt* 11637 (EA; K; LISC; MAL; P; PRE; SRGH; UPS). **Mozambique.** M: Namaacha, 30.4 kms. from Port Henrique, fl. & fr., 18.vii.1971, *Marques* 2312 (BM; LMU; SRGH).
     From Cameroon to Ethiopia and around Congolian forests to Angola, Namibia and S. Africa (Transvaal). Permanent or semi-permanent pools and rivers; up to 1750 m.

2. **Nymphoides indica** (L.) O. Kuntze, Rev. Gen. Pl. **2**: 429 (1891). —Taylor in F.W.T.A., ed. 2, **2**: 302 (1963). —Marais & Verdoorn in Fl. S. Afr. **26**: 243 (1963) pro minore parte. —Friedrich-Holzhammer in Merxm., Prodr. Fl. SW. Afr. **111**: 1 (1967) pro minore parte. —A. Raynal in Mitt. Bot. Staatss. München **10**: 126, map 3 (1971); in Adansonia, Sér. 2, **14**, 3: 416 (1974). Types from India and Sri Lanka.

Subsp. **occidentalis** A. Raynal in Adansonia, Sér. 2, **14**, 3: 8, t. 18 (1974); in Fl. Afr. Centr., Menyanthaceae: 12 t. 6 (1975). TAB. **17**. Type from Cameroon.

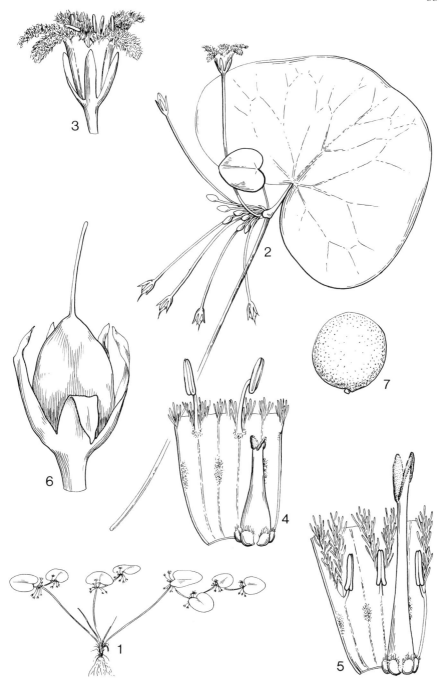

Tab. 17. NYMPHOIDES INDICA subsp. OCCIDENTALIS. 1, habit (× $\frac{1}{15}$); 2, inflorescence, flowers carried above water, capsules below it (× $\frac{2}{3}$), 1–2 from *Biegel & Russell* 3711; 3, flower (× $2\frac{1}{2}$); 4, corolla opened out to show pistil and stamens (× 5), 3–4 from *Bullock* 2162; 5, corolla opened out to show pistil and stamens (× 5), from *Biegel & Russell* 3711; 6, capsule with calyx (× 6), from *Symoens* 11196; 7, seed (× 6), from *Killick & Leistner* 3228.

Menyanthes indica L., Sp. Pl. **1**: 145 (1753). Type from India.
Limnanthemum indicum sensu Eyles in Trans. Roy. Soc. S. Afr. **5**, 4: 444 (1916).
Limnanthemum thunbergianum sensu Eyles, loc. cit. (1916) quoad *Allen* 82.

Perennial; stock short thick, stolons tufted, floating, numerous; stems and petioles 4–8 mm. thick in life. Petiole 1–6 cm; lamina 5–15(25) cm. in diam., circular. Flowers 10–30 heterostylous; pedicels 3.5–5(10) cm. long. Calyx 5–9 mm. long; lobes spreading at anthesis. Corolla tube yellow 7–9 mm. long, densely and uniformly hairy inside, ± twice as long as the tube. Capsule 4–5 mm. in diam., ovoid, shorter than the calyx, 10 or more seeded. Seeds 1.5–2.2 mm. long, lenticular ± tuberculate.

Caprivi Strip. Singalamwe, fl. & fr. 1.i.1959, *Killick & Leistner* 3228 (K; PRE; SRGH). **Botswana**. N: Mohombo Distr., alongside Okavango R., fl. & fr. 16.ii.1979, *Smith* 2661 (BM; MN; PRE; SRGH). **Zambia**. B: Kande Lake about 13 km. NE. of Mongu, fl. & fr. 9.i.1959, *Zinderen Bakker* 905 (PRE). N: Chinsali Distr., Shiwa Ngandu, fl. & fr. 17.i.1959, *Richards* 10714 (SRGH). W: Chingola, Kafue R., fl. & fr. 8.x.1954, *Fanshawe* 1596 (K). C: Lusaka Distr., Chongwe R., N. of Kasisi, fl. & fr. 29.ix.1972, *Strid* 2207 (K). S: Namwala, Lochinvar Nat. Park, fr. 31.v.1972, *Rees* 74 (SRGH). **Zimbabwe**. C: Ruwa R., next to Widdecombe Rd., fl. & fr. 2.ii.1966, *Howard-Williams* in GHS 181457 (K; SRGH). **Malawi**. S: Zomba Distr., Lake Chilwa, fr. 10.viii.1971, *Howard-Williams* 228 (SRGH). **Mozambique**. Z: Chinde, fr. 19.viii.1974, *Bond* W532 (SRGH). M: 7 kms. from Ponta do Ouro on the way to Zitundo, fr. 15.xi.1971, *Correia & Marques* 2106 (SRGH).

Widely distributed in tropical Africa, from Senegal to East Africa and south to Angola and Namibia. Lakes, rivers, swamps and pools; up to 1650 m.

The species as a whole is pantropical. It can tolerate considerable variation in water levels and can withstand temporary drying up of pools (when it may root at the nodes). It reproduces very effectively by vegetative means and can become a weed of waterways.

3. **Nymphoides thunbergiana** (Griseb.) Kuntze, Rev. Gen. Pl. **2**: 429 (1891). —A. Raynal in Mitt. Bot. Staatss. München **10**: 129, map 3 (1971); in Adansonia, Sér. 2, **14**, 3: 422, t. 19 (1974). —Gibson, Wild Fl. Natal: 80, t. 80, fig. 5 (1975). Type from S. Africa.
Limnanthemum thunbergianum Griseb., Gen. Sp. Gent.: 345 (1839). —Hill & Prain in F.C. **4**, 1: 1120 (1909). Type as above.
Nymphoides indica sensu Marais & Verdoorn in Fl. S. Afr. **26**: 243 (1963) pro majore parte.

Perennial; stock thick, floating stolons few. Petiole 2–6 cm.; lamina 3–10(20) cm. in diam., subcircular. Flowers 5–20 heterostylous, pedicels 2.5–6(8) cm. Calyx 4–6.5 mm. long, spreading at anthesis. Corolla pale yellow, tube 5–6 mm. long, lobes 1.2–1.5 mm. or more long, fimbriate laciniate along margins and midvein inside and sometimes densely hairy at the throat. Capsule ± 5 mm. in diam., ovoid, ± as long as the calyx, usually less than 10 seeded. Seeds 1.6–2.2 mm. long, lenticular ± tuberculate.

Botswana. N: Thamalakane R. near its confluence with Boro R., fl. 22.i.1972, *Biegel & Gibbs-Russell* 2711 (K; SRGH). **Zambia**. W: Kitwe, fr. 24.x.1957, *Fanshawe* 3808 (K). **Zimbabwe**. C: Harare, Hatfield, fr. 8.v.1934, *Gilliland* 69 (K). E: Chimanimani Mts., Bundi R., fl. & fr. 26.x.1959, *Goodier & Phipps* 288 (K; SRGH).

Also known from S. Africa and Madagascar. Pools and rivers; up to 1600 m.

4. **Nymphoides forbesiana** (Griseb.) Kuntze, Rev. Gen. Pl. **2**: 429 (1891). —A. Raynal in Mitt. Bot. Staatss. München **10**: 128, map 4 (1971); in Adansonia, Sér. 2, **14**: 424, t. 20 (1974); in Fl. Afr. Centr., Menyanthaceae: 6, t. 3: Type: Mozambique, *Forbes* s.n. (K, lectotype), lectotype chosen by A. Raynal loc. cit.
Limnanthemum forbesianum Griseb. Gen. Spl. Gent.: 345 (1839) pro parte.
Limnanthemum niloticum Kotschy & Peyr., Pl. Tinn.: 28 t. 9A (1867). —N.E. Br. in F.T.A. **4**, 1: 585 (1904). Type from Sudan.
Limnanthemum thunbergianum sensu N.E. Br., loc. cit.: 584 (1904) pro majore parte.
Limnanthemum kirkii N.E. Br. loc. cit.: 585 (1904). Types from Kenya and Tanzania.
Limnanthemum whytei N.E. Br. loc. cit.: 585 (1904). Type from Kenya.
Nymphoides indica sensu Eyles in Trans. Roy. Soc. S. Afr. **5**, 4: 444 (1916) quoad *Swynnerton* 393. —Taylor in F.W.T.A., ed. 2, **2**: 302 (1963) pro parte.

Perennial, sometimes annual; floating stolons long and thin, 1–3 mm. thick in life. Petiole 0.5–1(3) cm., lamina 12 cm. in diam., subcircular with a deep narrow sinus. Flowers 10–20 heterostylous; pedicels 3–6 cm. long. Calyx 3–4 mm. long, spreading at anthesis. Corolla golden yellow 10–15 mm. long, tube 3–3.5 mm. long, lobes 6–7 mm. long, fimbriate-laciniate along margins and midvein inside, hairy, sometimes more densely so at the throat. Capsule ovoid, 4–5 × 5.7 mm. slightly longer than the calyx, up to 12 seeded. Many fruits abort. Seeds 1.6–2.2 mm, round, smooth to spiny or glochidiate.

Botswana. N: Mahu Pan, fr. 29.iii.1973, *Smith* 495 (SRGH). **Zambia**. B: Masese, fr. 3.v.1961, *Fanshawe* 6533 (K). W: Mwinilunga, fr. 8.vi.1963, *Loveridge* 855 (K; SRGH). C: Mkushi R. fr. xii.1960, *Richardson & Livingstone* s.n. (K). **Zimbabwe**. W: Shangani Reserve, fr. 1.iv.1951, *West* 3154 (K). C: Chegutu Distr., Poole Farm, fl. & fr. 6.iv.1952, *Hornby* 3296 (K; SRGH). E: Gungunyana Forest, Mt. Silinda, fr. xii.1937, *Obermeyer* s.n. (PRE). S: Beitbridge Distr., near Chiturupadzi Store, 40 km. NNW. of the Bubi-Limpopo R. confluence, fl. & fr. 12.v.1985, *Drummond* 5765 (K; SRGH). **Mozambique**. N: Niassa Prov. about 25 km. along the road from Marrupa to Lichinga, fl. & fr. 10.viii.1981, *Jansen, de Koning & de Wilde* 160 (LMU). GI: Gaza, Limpopo, fl. & fr. 27.vii.1973, *Correia & Marques* 3087 (LMU). M: Namaacha, fl. & fr. 25.viii.1967, *Sousa* 4943 (PRE).
Widespread in tropical Africa, from Ivory Coast to Sudan, south to Angola and S. Africa. Ponds and rivers; up to 1660 m.

5. **Nymphoides rautanenii** (N.E. Br.) A. Raynal in Mitt. Bot. Staatss. München **10**: 124, map 1 (1971); in Adansonia, Sér. 2, **14**, 3: 434, t. 25 (1974); in Fl. Afr. Centr., Menyanthaceae: 10, t. 5 (1975). Type from Namibia.
   Limnanthemum rautanenii N.E. Br. in F.T.A. **4**, 1: 585 (1904) as "rautaneni". Type as above.
   Nymphoides indica sensu Marais & Verdoorn in Fl. S. Afr. **26**: 243 (1963) pro parte. — Friedrich-Holzhammer in Merxm., Pr. Fl. SW. Afr. **111**: 1 (1967) pro parte.

Annual sometimes perennial; floating stolons numerous; stems and petioles 1–3 mm. thick in life. Petiole 3–6 cm. long; lamina 3–5(11) cm. in diam., circular. Flowers (5)10–15 homostylous; pedicels 0.5–2.5(3) cm. long, slender. Calyx c. 3 mm. long. Corolla yellow less than 10 mm. long, lobes villous at margins and along midvein inside, tube 1.5–2 mm. long, lobes 4–7 mm. long. Capsule 5–7 mm. in diam., globose, exceeding the calyx, 10–15 seeded. Seeds 1.3–1.6 mm. long, lenticular, finely lenticular or tuberculate.

Botswana. N: 5 km. N. of Shakawe, fl. & fr. 28.iv.1975, *Gibbs-Russell* 2845 (K; SRGH). **Zambia**. B: Sesheke, fl. 24.i.1924, *Boile* s.n. (PRE). S: Kafue Basin, fr. 28.iv.1964, *van Rensburg* 2902 (K; SRGH). **Zimbabwe**. W: Hwange, fl. & fr. iv.1932, *Levy* 26 (PRE). **Mozambique**. GI/M: Guija Distr., along Limpopo R. fr. vii.1915, *Gazaland Exped.* 15811 (PRE).
Also known from Zäire (Shaba), Angola and Namibia. Temporary pools; up to 1100 m.

6. **Nymphoides milnei** A. Raynal in Mitt. Bot. Staatss. München **10**: 132, map 5 (1971); in Adansonia, Sér. 2, **14**, 3: 428, t. 22 (1974). Type: Zambia, Mwinilunga Distr., Matonchi Farm, *Milne-Redhead* 4317 (K, holotype).

Small herb, probably annual without floating stolons. Petiole 2–3 mm. long, membraneous; lamina 20–32 × 15–27 mm., mostly ovate with a short gaping sinus 0.8–1.2 cm. long. Flowers few c. 4, heterostylous; pedicels c. 20 mm. long. Calyx c. 3 mm. long. Corolla yellow 7–8 mm. long, tube c. 3 mm. long, lobes c. 4 mm. long. Capsule c. 3 × 2.7 mm., obovoid, 3–4 or less seeded. Seeds 1.6 × 0.9 mm., lenticular, with strongly convex faces.

Zambia. W: c. 1 km. SW. of Matonchi Farm, fl. & fr. 24.i.1938, *Milne-Redhead* 4317 (K).
Known only from the type gathering. Temporary pool.

7. **Nymphoides tenuissima** A. Raynal in Mitt. Bot. Staatss. München **10**: 131, map 5 (1971); in Adansonia, Sér. 2, **14**, 3: 438, t. 27 (1974); in Fl. Afr. Centr., Menyanthaceae: 4, t. 2 (1975). TAB. **18**. Type: Zambia, Kaputa Distr., Mweru-Wantipa, Bulaya, *Richards* 9175 (K, holotype).

Small annual without stolons. Petiole short, less than half as long as lamina. Lamina 2.4–5.8 × 2.1–4.7 cm., mostly ovate with an open sinus. Flowers 7–10(15), homostylous; pedicels 7–8(15) mm. long. Calyx c. 3 mm. long appressed to the corolla. Corolla white, 5–7 mm. long, lobes fimbriate villous on edges and midvein. Capsule globose, just exceeding the calyx, (5)10–15 seeded. Seeds 1.3–1.6 mm. long, broadly elliptic-ovate, flattened, often finely tuberculate.

Zambia. N: Kaputa Distr., Mweru-Wantipa, Rd. to Bulaya, fl. & fr. 12.iv.1957, *Richards* 9175 (K). W: 7 km. E. of Chizela, fl. & fr. 27.iii.1961, *Drummond & Rutherford-Smith* 7443 (K; SRGH).
Also known from Zaire (Shaba). Temporary pools; c. 900–1200 m.

56

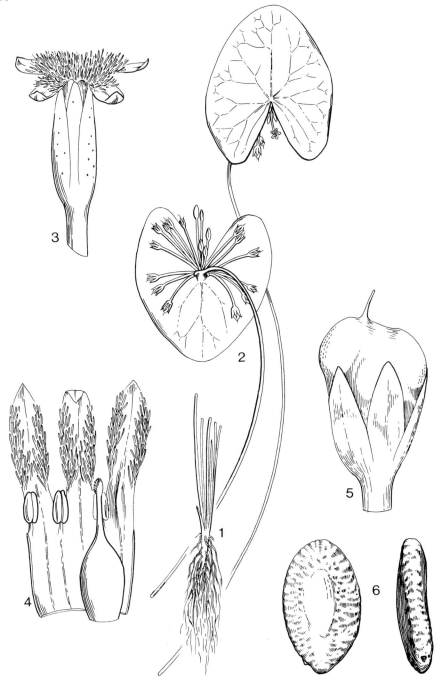

Tab. 18. NYMPHOIDES TENUISSIMA. 1, base of plant (× ⅔), from *Richards* 9175; 2, leaves with inflorescence, ventral and dorsal surfaces, capsules carried below water surface (× ⅔), from *Drummond & Rutherford-Smith* 7443; 3, flower (× 10); 4, corolla opened out to show pistil and stamens (× 15); 5, capsule (× 10); 6, seed in dorsal and lateral views (× 20), 3–6 from *Richards* 9175.

115. HYDROPHYLLACEAE

By B. Verdcourt

Annual or perennial herbs, rarely subshrubs, often scabrid. Leaves usually alternate or radical, exstipulate, entire to deeply pinnately or palmately lobed, with margins entire, serrate or lobulate. Flowers regular, hermaphrodite, in scorpioid or trichotomous cymes, false racemes or clustered, less often solitary. Calyx deeply divided into 5 rarely 8–12 imbricate or open equal or unequal lobes or 5 with appendages between. Corolla often blue, variously shaped, the lobes usually 5, rarely 8–12, imbricate or rarely contorted. Stamens as many as the corolla lobes, inserted at the sinuses or low in the tube; filaments filiform or dilated at the base; anthers bilocular, versatile, opening by longitudinal slits, included or exserted. Disk inconspicuous or absent. Ovary superior (very unusually inferior) unilocular with 2 parietal fleshy placentas or 2(3)-locular. Ovules 2 or more on each placenta; style filiform, usually bifid or styles 2; stigmas simple or capitate. Fruit usually a loculicidal or less often septicidal bivalved or irregularly splitting capsule. Seeds oblong, globose or angled with tuberculate reticulate or rugose testa; albumen copious and fleshy; embryo small; cotyledons entire.

About 18 genera with c. 300 species, mainly American; only one native species. *Phacelia tanacetifolia* Benth. has been cultivated in East Africa but I have seen no material from the Flora Zambesiaca area. *Wigandia* is extensively grown in upland tropical Africa but I have only seen one specimen from the Flora Zambesiaca area namely *Wigandia urens* (Ruiz & Pavon) Kunth var. *caracasana* (Kunth) D.N. Gibson f. *africana* (Brand) Verdc. —*Biegel* 2626 (K; SRGH) (Zimbabwe, Harare fl. 19.vii.1968). The collector describes it as a coarse shrub 3.6 m. tall and wide including branches, inferior surface of the leaves and inflorescences viscous-hairy and flower deep purple with white throat and tube. *Nemophila* species are probably also grown. These genera can be separated as follows

1. Slender herbs of swamps etc.; styles 2; plants native - - - - - - **Hydrolea**
 – Slender herbs to subshrubs; style(s) 1 or 2; plants cultivated - - - - - 2
2. Large subshrubby herb or shrub to 4 m., with huge serrate or crenate but not deeply divided leaves; styles 2 - - - - - - - - - - - - - **Wigandia**
 – Smaller herbs, with more deeply divided leaves, style 1, bifid - - - - 3
3. Corolla tubular; flowers in cymes - - - - - - - - - - **Phacelia**
 – Corolla rotate (in species likely to be grown in Africa); flowers usually solitary - - - - - - - - - - - - - - - **Nemophila**

HYDROLEA L.

Hydrolea L., Sp. Pl. ed. 2: 328 (1762); Gen. Pl. ed. 6: 124 (1764). —A.W. Bennett in Journ. Linn. Soc. Bot. **11**: 266–279, t. 1 (1870). —Brand in Engl., Pflanzenr. IV 251, **59**: 174–185 (1913) —Davenport in Rhodora **90**: 169–208 (1988) nom. conserv.*

Herbs or subshrubs, sometimes spiny, glabrous or glandular-pubescent. Leaves alternate, entire. Flowers in peduncled bracteate cymes, forming panicles or in false racemes or axillary clusters. Calyx lobes 5, imbricate at the base, open above. Corolla often blue, rotate-campanulate, 5-lobed. Stamens 5, inserted at the sinuses of the corolla; filaments filiform, usually dilated at base; anthers sagittate. Ovary 2(3)-locular with fleshy placentas adnate to the septa; ovules many per locule. Styles 2, subulate, distinct from the base; stigmas simple or capitate. Capsule globose, ellipsoid or ovoid, thin, usually septicidally bivalved; seeds numerous, minute.

About 20 species in the tropics and subtropics of both Old and New Worlds; 7 occur in Africa but only one very locally in the Flora Zambesiaca Area.

Hydrolea brevistyla Verdc. in F.T.E.A., Hydrophyllaceae: 4 (1988). TAB. **19**. Type: Zambia, Mapanza, Simasunda, *Robinson* 2820 (BR; K, holotype; LISC).

Weak erect annual herb 7.5–30 cm. tall; stems ± compressed at the base, almost winged

---

* The reasons for conservation are not evident.

58

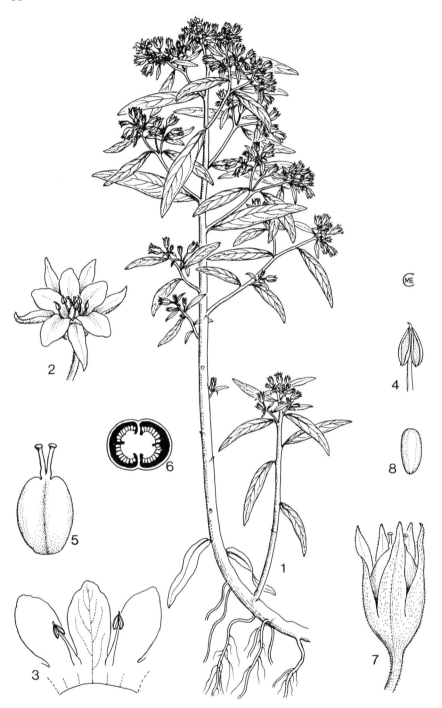

Tab. 19. HYDROLEA BREVISTYLA. 1, habit (×⅔); 2, flower (× 4); 3, part of corolla opened out to show stamens (× 8); 4, stamen (× 20); 5, pistil (× 12); 6, ovary in transverse section (× 14); 7, capsule with calyx (× 6); 8, seed (× 40), all from *Robinson* 5200.

in the dry state, decumbent below in some stouter specimens and rooting at the nodes, with a few minute hairs on very young parts. Leaves 0.1–4.5 × 0.1–0.9 cm., lanceolate or elliptic to linear-oblong, bluntly acute at the apex, cuneate at the base, pale green, glabrous; petiole obsolete or very short. Cymes terminal, few to many-flowered; pedicels 2–4 mm. long. Calyx micropuberulous; tube 1.2 mm. long; lobes 3.5–5 × 0.15 × 1.5 mm. ovate-lanceolate, minutely puberulous. Corolla deep blue, white at base, tube 0.8–1.3 mm. long; lobes 2–2.5 × 1.3 mm. oblong-elliptic. Anthers apiculate 0.5–0.8 mm. long; filaments 1 mm. long. Ovary 1.5 mm. long, the style and stigmas together 1 mm. long. Fruit 3.5 × 2.5 mm., cylindrical-ellipsoid.

Zambia. N: Chishimba Falls, fl. & fr. 20.v.1962, *Robinson* 5200 (K; LISC). S: 4.8 km. E. of Mapanza, Simasunda Dambo, fl. 4.iv.1954, *Robinson* 660 (K).

Edges of dambos beginning to dry out or just dried, also on horizontal slabs of sandstone seasonally irrigated by overflow from river; 1050 m.

116. BORAGINACEAE

By E.S. Martins
(*Trichodesma* by R.K. Brummitt)

Herbs, shrubs or rarely trees. Leaves alternate, rarely subopposite or opposite, simple exstipulate. Inflorescence usually cymose, composed of one or more helicoid or scorpioid cymes, sometimes paniculate or racemose or sometimes flowers solitary, terminal or axillary. Flowers ⚥ or sometimes unisexual, actinomorphic or sometimes zygomorphic. Calyx (4)5(6)-lobed or 3–5-toothed, imbricate or valvate. Corolla (4)5(6)-lobed, imbricate or contorted in bud, tubular, funnel-shaped, campanulate, salver-shaped or rotate; tube sometimes with folds, scales or hairs in the throat. Stamens (or staminodes) as many as corolla lobes and alternate with them, inserted on the corolla tube; anthers 2-thecous, dehiscing longitudinally, introrse, dorsifixed. Disk present or absent. Ovary superior, entire or deeply 4-lobed, bi-locular or spuriously 4-locular, placentation axial, ovules 4, erect or spreading; style 1, terminal or gynobasic, entire or cleft or twice cleft, or styles 2, terminal (*Coldenia*). Fruit drupaceous with a 1–4-seeded stone or splitting into 4 pyrenes, or of one to four-seeded nutlets; seeds generally without endosperm.

A widely-distributed family mainly in temperate and tropical regions of the Old and New World.

In addition to the native and introduced species included in this account, some species occur as weeds or are cultivated in the Flora Zambesiaca area either as decorative plants or as fodder, namely *Cordia alliodora* (Ruiz & Pavon) Oken, *Cordia obliqua* Willd., *Heliotropium amplexicaule* Vahl, *Heliotropium arborescens* L., *Borago officinalis* L., *Amsinckia calycina* (Moris) Chater, *Echium plantagineum* L., *Echium vulgare* L., *Symphytum officinale* L. and *Symphytum x uplandicum* Nyman.

Key to the genera

1. Trees or shrubs; style terminal   -   -   -   -   -   -   -   -   -   -   2
   – Herbs or subshrubs; style terminal or gynobasic   -   -   -   -   -   -   5
2. Fruit dry, divisible into two 2-seeded mericarps; style almost absent; stigma 1, ring-like,
      subterminal   -   -   -   -   -   -   -   -   -   -   -   **5. Argusia**
   – Fruit fleshy; style well-developed; stigmas 2 or 4, or 1 bilobed, not as above   -   -   3
3. Style with each branch again divided; stigmas 4, slender or clavate   -   -   - **1. Cordia**
   – Style cleft or subentire; stigmas 2, subcapitate or peltate, or 1 bilobed   -   -   -   4
4. Calyx shortly lobed or dentate; corolla lobes rounded; endocarp with thin longitudinal
      lamellas   -   -   -   -   -   -   -   -   -   -   -   -   **2. Bourreria**
   – Calyx lobed to near base; corolla lobes oblong or lanceolate; endocarp with small surface
      cavities   -   -   -   -   -   -   -   -   -   -   -   **3. Ehretia**
5. Ovary entire or subentire; style terminal   -   -   -   -   -   -   -   -   6
   – Ovary deeply lobed; style gynobasic   -   -   -   -   -   -   -   -   -   7
6. Styles 2; stigmas capitate; flowers 4-merous, solitary, extra-axillary   -   - **4. Coldenia**
   – Style 1; stigma ring-like, below the apex of style; flowers 5-merous, in scorpioid spikes or
      solitary   -   -   -   -   -   -   -   -   -   -   -   **6. Heliotropium**

7. Corolla throat naked; connectives prolonged above the anthers by lamellate
   appendages   -   -   -   -   -   -   -   -   -   -   -   -   -   -   -   8
 – Corolla throat with folds or fornices; connectives not prolonged above the
   anthers   -   -   -   -   -   -   -   -   -   -   -   -   -   -   -   9
8. Appendages of the anthers twisted; gynobase pyramidal; calyx strongly accrescent in
   fruit   -   -   -   -   -   -   -   -   -   -   -   -   -   **7. Trichodesma**
 – Appendages of the anthers straight; gynobase flat; calyx not or only little accrescent in
   fruit   -   -   -   -   -   -   -   -   -   -   -   -   **8. Cystostemon**
9. Nutlets ovoid (circular in cross-section) erect, smooth, attached to the gynobase by the base;
   cymes conspicuously bracteate   -   -   -   -   -   -   -   **10. Lithospermum**
 – Nutlets ovoid-depressed (elliptic in cross-section), covered with glochidiate spines, attached to
   the gynobase by the upper part; cymes not bracteate   -   -   -   -   **9. Cynoglossum**

1. CORDIA L.

Cordia L., Sp. Pl.: 190 (1753); Gen. Pl. ed. 5: 87 (1754).

Trees or shrubs sometimes dioecious. Leaves alternate, rarely subopposite, petiolate, entire to coarsely toothed. Cymes arranged in lax or dense or sometimes very contracted and subglobose panicles, ebracteate, terminal or axillary. Flowers ⚥ or unisexual, actinomorphic, pedicelled or subsessile, the pedicels joined. Calyx tubular or ± campanulate, sometimes sulcate, usually splitting irregularly, 3–5-toothed, enlarging in fruit. Corolla 4–5(7)-lobed, funnel-shaped to salver-shaped, white or yellowish; lobes imbricate or subcontorted in bud, shallow and obscure or oblong and conspicuous, patent or recurved at anthesis. Stamens or staminodes inserted in the corolla tube, as many as the corolla lobes and alternate with them; anthers oblong. Ovary entire, 4-locular with 1 ovule in each locule (or reduced and abortive in ♂ flowers); style terminal, twice cleft, with stigmatic branches linear or clavate. Fruit drupaceous, partially or wholly surrounded by the enlarged persistent calyx; stone with 1–4 exalbuminous seeds, cotyledons plicate

A genus of about 300 species distributed in warm and tropical regions of the world, with a strong concentration of species in the American Continent.

1. Calyx in flower with 10–12 well-marked ridges, tomentose outside; corolla lobes shallow, much
   broader than long; leaves farinose below, with or without spreading hairs   1. *africana*
 – Calyx in flower smooth or with not well-marked ridges; corolla lobes rounded or longer than
   broad; leaves usually not farinose   -   -   -   -   -   -   -   -   2
2. Corolla more than 30 mm. long, funnel-shaped; fruit wholly enclosed in the enlarged urceolate
   calyx; plants from coral beaches   -   -   -   -   -   -   -   2. *subcordata*
 – Corolla less than 25 mm. long, cylindrical to narrowly funnel-shaped; fruit not as above; plants
   from inland   -   -   -   -   -   -   -   -   -   -   -   -   -   3
3. Leaves glabrous at maturity   -   -   -   -   -   -   -   -   -   -   -   4
 – Leaves not glabrous at maturity   -   -   -   -   -   -   -   -   -   -   7
4. Leaves entire   -   -   -   -   -   -   -   -   -   -   -   -   -   5
 – Leaves subentire to toothed, crenate or serrate   -   -   -   -   -   -   6
5. Petiole 1.3–3.0(4.0) cm. long; leaf lamina up to 2 times as long as wide   -   -   3. *goetzei*
 – Petiole 0.5–1.5 cm. long; leaf lamina always more than 2 times, often more than 3 times as long
   wide   -   -   -   -   -   -   -   -   -   -   -   -   -   4. *mukuensis*
6. Panicles pubescent to glabrous, lax; leaves crenate-serrate, petiole very
   slender   -   -   -   -   -   -   -   -   -   -   -   -   -   13. *caffra*
 – Panicles pilose to hispid, contracted; leaves ± undulate, remotely crenate or minutely lobed,
   petiole not very slender   -   -   -   -   -   -   -   -   -   -   15. *sp. B*
7. Lower surface of the leaves grey-yellowish, minutely woolly, the hairs bi-branched,
   medifixed   -   -   -   -   -   -   -   -   -   -   -   -   -   11. *torrei*
 – Lower surface of the leaves not woolly, the hairs, simple, stellate or dendriform, not
   bi-branched   -   -   -   -   -   -   -   -   -   -   -   -   -   8
8. Leaf margins denticulate to coarsely dentate, each tooth ending in a point up to 1.5 mm. long;
   both surfaces with scattered short hairs   -   -   -   -   -   -   -   9
 – Leaf margins and surfaces not as above   -   -   -   -   -   -   -   11
9. Cymes in very contracted 1.0–2.5 cm. long panicles; fruiting calyx widely campanulate, very
   narrowed at base; leaves cuneate at base   -   -   -   -   -   -   14. *sp. A*
 – Cymes in narrow, slender, long-peduncled panicles; leaves cuneate to rounded
   at base   -   -   -   -   -   -   -   -   -   -   -   -   -   10

10. Corolla tube 6–7 mm. long, densely hairy at the mouth; corolla lobes 4, oblong-lanceolate; leaves lanceolate to oblanceolate - - - - - - - - - - - 7. *stuhlmannii*
– Corolla tube 4.5–5.5 mm. long, not much densely hairy at the mouth; corolla lobes 5, oblanceolate to obovate; leaves ovate to obovate - - - - - 8. *mandimbana*
11. Leaves opposite to alternate, narrowly obovate to elliptic, above rough and often densely covered with cystoliths, below with short ± appressed hairs in all directions and longer hairs in the axils and along the midrib - - - - - - - - - - 12. *sinensis*
– Leaves without the above combined characters - - - - - - - - - - 12
12. Panicles small, contracted, pilose to hispid; leaves elliptic to oblong-elliptic, glabrous above, subglabrous below - - - - - - - - - - - - 15. *sp. B*
– Plants without the above combined characters - - - - - - - - - 13
13. Calyx ridged or prominently nerved; style 10–22 mm. in total length; fertile anthers 2–4 mm. long, filaments 4–8 mm. long - - - - - - - - - - - - 14
– Calyx neither ridged nor prominently nerved; style 7–9 mm. in total length; fertile anthers 1.5–2.0 mm. long, filaments 1.5–4.0 mm. long - - - - - - - - - 15
14. Fruit up to 40 × 35 mm., with a horn-shaped point; fruiting calyx up to 35 mm. wide at the mouth, very prominently nerved, margin revolute; corolla tube 12–17 mm. long, puberulous outside; inflorescence minutely tomentose - - - - - - - - - 10. *grandicalyx*
– Fruit up to 27 × 16 mm., without horn-shaped point; fruiting calyx up to 25 mm. wide at the mouth, prominently nerved, margin revolute but soon eroded; corolla tube 6–10 mm. long, glabrous outside; inflorescence (including calyx) densely villous - - - - 9. *pilosissima*
15. Calyx glabrous or subglabrous outside; stigmatic branches flattened, subfoliaceous, ± deeply lobed; leaves 3–7-nerved from the base, glabrous above or only hairy on the nerves; leaf scars much prominent - - - - - - - - - - - - - 5. *myxa*
– Calyx densely scabrous (on ♀ & ☿ flowers) or roughish and tawny-tomentose (on ♂ flowers) outside; stigmatic branches linear, not lobed; leaves pinnately nerved or 3-nerved from base, densely covered with short, rigid, erect excrescences above; leaf scars a little prominent - - - - - - - - - - - - - - 6. *monoica*

1. **Cordia africana** Lam., Tab. Encycl. Méth. Bot. **1**: 420 (1792). —Heine in Hutch. & Dalz., F.W.T.A. ed. 2, **2**: 320, t. 276 (1963). —Taton in Fl. Congo, Rwanda et Burundi, Boraginaceae: 4 (1971). —Jansen in Belmontia n.ser., **12**: 170, t. 19 & ph. 30 (1981). Type from Ethiopia.
    Cordia abyssinica R. Br. [in Salt, Abyss. App. IV: 64 (1814)] ex A. Rich., Tent. Fl. Abyss. **2**: 80 (1851). —Baker & Wright in F.T.A. **4**, 2: 8 (1905). —Brenan & Greenway, T.T.C.L.: 75 (1949). —Williamson, Useful Pl. Nyasal.: 40 (1956). —Burtt Davy & Hoyle, N.C.L. ed. 2: 32 (1958). —F. White, F.F.N.R.: 364 (1962). —Palmer & Pitman, Trees of Southern Afr. **3**: 1935 cum tab. (1973). —R.B. Drumm. in Kirkia **10**: 271 (1975). —Palgrave, Trees of Southern Afr.: 799 cum tab. (1977). Type from Ethiopia.

Small to medium sized deciduous tree, up to 24 m. high; trunk usually forking a few meters from the base; bark grey to dark brown, shallow fissured; crown spreading, dome-shaped; branchlets glabrous, powdery or pubescent, sometimes hairy. Leaves alternate or sometimes almost ternate; petiole (1.5)2.5–10.5 cm. long, canaliculate, glabrous or tawny-powdery and sometimes also tawny-pubescent; lamina 6–21 × 4.0–16.5 cm., ovate or broadly ovate, sometimes subcircular or elliptic, glabrous or on the youngest leaves scabrous or minutely scaly above, powdery and often tawny-hairy on the nerves and veins below, usually rounded and suddenly shortly acuminate, sometimes acute at the apex, subacute to cordate and often asymmetrical at the base, with entire or shallowly crenate margins, usually stiffly coriaceous, with 5–7 secondary nerves on each side of the midrib, tertiary nerves perpendicular to the secondary ones, both prominent below. Cymes arranged in terminal leafy usually ample panicles; rhachis and branches tawny or brown, minutely tomentose. Flowers ☿, on pedicels up to 1 mm. long. Calyx 7–9 mm. long and 4–7 mm. wide at the mouth, tubular, with 10–12 well marked ridges, coriaceous, minutely tomentose outside, glabrous inside, opening by an operculum and later splitting into 3–5 teeth, sometimes becoming bilabiate. Corolla 17–21(24) mm. long, funnel-shaped, white; tube 15–19(22) mm. long; lobes 1.5–3.0 × 10–13(15) mm., retuse and with a pubescent mucron c. 1 mm. long, margin undulate. Stamens enclosed, inserted at 3–5 mm. from the base of the corolla; anthers 2.0–2.5 mm. long; filaments 7–10 mm. long, slender, with a few pellucid hairs at the base. Ovary c. 2.0 × 1.5 mm., ovoid, glabrous; style 13–18 mm. long, first forked at 7–12 mm. from the base, stigmatic branches c. 1.5 mm. long, more or less clavate. Fruit c. 12 × 8 mm., ovoid, ellipsoid or obovoid, apiculate, glabrous, dark brown, surrounded at base by the widened cup-shaped calyx; pyrene subquadrangular in cross section. Seeds 1–2, very rarely 3–4.

Zambia. N: Mbala Distr., Lunzua Falls Forest, fl. 24.v.1967, *Richards* 22261 (K). W: Ndola, fl. 21.iii.1954, *Fanshawe* 995 (K). E: Chipata Township, fl. 25.iv.1952, *F. White* 2458 (FHO; K) (planted). S. Livingstone Distr., fl. iii.1953, *Seale* in GHS 42651 (SRGH). **Zimbabwe**. N: Bindura, Pote Valley, fl.

iv.1933, *Harvie* 46/33 (FHO; SRGH). W: Bulawayo, fl. x.1953, *Orpen* 46/53 (SRGH) (cultivated). E: Chipinge Distr., N.W. Farfell Farm, adjoining Gungunyana For. Res., fl. vi.1962, *Goldsmith* 150/62 (BM; COI; K; LISC; SRGH). S: Masvingo Distr., Great Zimbabwe, fr. 8.vii.1972, *Chiparawasha* 496 (K; SRGH). **Malawi**. N: Vizuzu Hill, on Rd. from Mzuzu, fl. 15.vi.1964, *Chapman* 2248 (FHO). C: Chipata Mt., Nkhota Kota Hills, st. 22.xi.1962, *Chapman* 1753 (SRGH). S: Thyolo Distr., Cholomwani For. Res., fr. 26.x.1960, *Chapman* 1026 (FHO; SRGH). **Mozambique**. N: Sanga, Mt. Chicungulo, near Malulo, fl. 2.iii.1964, *Torre & Paiva* 10975 (C; COI; FHO; LD; LISC; LMA; M; PRE; WAG). Z: Milange, Mt. Tumbini, fr. 30.vii.1949, *Andrada* 1805 (COI; LISC). MS: Mossurize, near Espungabera, fr. 30.x.1944, *Mendonça* 2677 (BR; J; LISC; LMU; MO; P; SRGH).

Widespread in tropical Africa from Guinea to Ethiopia and from the Sudan to Angola and S. Africa (Northern Transvaal); also recorded from tropical Arabia. Occasionally in dry forest, in dry evergreen forest, in mountain rainfall woodland and on rocky mountain slopes; often planted as an ornamental tree; 750–1700 m.

2. **Cordia subcordata** Lam., Tab. Encycl. Méth. Bot. **1**: 421 (1792). —Poir., Encycl. Méth. Bot. **7**: 41 (1806). —Baker, Fl. Maurit.: 200 (1877). —Baker & Wright in F.T.A. **4**, 2: 10 (1905). —Fosberg & Renvoize, Fl. Aldabra: 194, t. 31 fig. 1–3 (1980). Type from the Seychelles (Praslin I.).

Shrub up to 4 m. high (or a tree up to 15 m. high outside the Flora Zambesiaca area); branches angular when young, glabrous or subglabrous. Leaves alternate; petiole 2.5–7.0 cm. long, subglabrous; lamina 6–20 × 5–16 cm., ovate or widely ovate to circular, glabrous or with some minute appressed not bulbous-based hairs, older ones frequently dotted with groups of cystoliths above, more or less pubescent along the midrib and some secondary nerves and usually minutely tomentose in the axils of the principal secondary and tertiary nerves below, obtuse to rounded and often shortly acuminate at apex, obtuse to truncate, rarely subcordate at the base, with entire or slightly repand margins, papyraceous, with 6 pairs of secondary nerves. Inflorescence cymose, loosely branched, short, 6–20-flowered, terminal, often apparently opposite-leaved. Flowers ♀, heterostylous, on pedicels 2–8 mm. long joined near the apex. Calyx 12–15 mm. long and 4–5 mm. in diam. at the mouth, cylindric or slightly conical, coriaceous, usually with 3–5 rounded teeth up to 4 mm. long, glabrous or very sparsely strigose outside and inside but densely strigose inside on the teeth. Corolla 30–45 mm. long, infundibuliform, orange, glabrous or nearly so; limb nearly as broad as the length of the corolla; lobes 5–7, broad, rounded, sometimes with crenate margin, spreading. Stamens inserted at c. 20 mm. from the base of the corolla tube; filaments in the short-styled flowers c. 8 mm. long, in the long-styled flowers 3–5 mm. long, glabrous; anthers 2.5–3.5 mm. long. Ovary c. 3 mm. long, conical, glabrous; style in the short-styled flowers c. 22 mm. long, first-branched at c. 17 mm. and with stigmatic branches c. 1.5 mm. long, in the long-styled flowers c. 30 mm. long, first-branched at 21–23 mm. and with stigmatic branches 2.5–3.0 mm. long, flattened. Fruit a nut, c. 25 × 22 mm., subglobose, smooth, shining, completely enclosed in the enlarged urceolate calyx; mesocarp suberose; pyrene angular, much ridged and scabrous, containing 1–2 seeds.

Mozambique. N: I. de Mafamede, st. 8.x.1965, *Gomes e Sousa* 4857 (K; LMA). GI: I. de Santa Carolina, fl. & fr. 3–5.xi.1958, *Mogg* 28785 (K; PRE).
Known from the tropical coasts of east Africa through the islands of the Indian and Pacific Oceans to India, Indo-China and Polynesia. On coral-beaches.

3. **Cordia goetzei** Gürke in Engl., Bot. Jahrb. **28**: 307 (1900). —Baker & Wright in F.T.A. **4**, 2: 14 (1905). —Brenan & Greenway, T.T.C.L.: 76 (1949). —Gomes e Sousa, Dendrol. Moçamb. Estudo Geral **2**: 657, t. 211 (1967). —Fanshawe in Kirkia **6**: 176 (1968). —Drumm. in Kirkia **10**: 271 (1975). —Palgrave, Trees of Southern Afr.: 800, cum tab. (1977). —Hall-Martin & Drumm. in Kirkia **12**: 177 (1980). TAB. **20**. Type from Tanganyika.
    Cordia sp. — Schinz in Denkschr. Akad. Wiss. Wien, Math.-Nat. Kl. **78**: 436 (1905).
    Cordia myxa sensu Baker & Wright, loc. cit. (1905) excl. syn.

Small tree up to 8(10) m. high or sometimes a scandent shrub, dioecious, with hard timber; trunk and branches irregularly fluted, often square in section and even buttressed; bark smooth, grey-green in small specimens, becoming blue-grey and peeling away into thin plate-like flakes; branchlets glabrous, very rarely hairy. Leaves alternate; petiole 0.8–3.0(4.0) cm. long, canaliculate, glabrous; lamina 5–12 × 2.5–7.0 cm., elliptic or broadly elliptic to obovate, glabrous on the mature leaves, acute to obtuse at apex, cuneate to truncate and often asymmetrical at base, with entire margins, membranous to almost coriaceous, with 5–7 secondary nerves on each side of the midrib, not conspicuous on the lower surface, reticulate. Cymes arranged in shortly peduncled, slender, terminal and

Tab. 20. CORDIA GOETZEI. 1, flowering twig (× ½); 2, flower (× 4); 3, gynoecium (× 4), 1–3 from *Gomes e Sousa* 4499; 4, fruit (× 1), from *Maĉedo* 1227.

pseudo-axillary pseudo-dichotomous lax 2.5–6.0 cm. long and 3–9 cm. wide panicles; rhachis and branches glabrous or with scattered hairs. Flowers ♂ or ♀, on pedicels up to 2 mm. long. ♂ flowers with buds c. 5 mm. long, obovoid, glabrous or subglabrous but often with a tuft of minute hairs at the apex; calyx 5–6 mm. long, tubular-campanulate, irregularly 3–5-toothed, membranous, glabrous or subglabrous outside, minutely strigose inside; corolla whitish, with tube 3.5–5.0 mm. long, cylindric, glabrous outside and inside save by a few hairs on the line of the stamens and with 4–5 lobes, 4.5–5.5 × 1.4–1.6 mm., oblong-lanceolate, acute to obtuse, ± reflexed, glabrous; stamens 4–5, inserted at the corolla-throat, filaments 3.5–5.0 mm. long, slender, hairy to the base, anthers 1.5–2.0 mm. long; ovary vestigial without style. ♀ flowers with buds as in ♂ flowers; calyx 4.5–5.5 mm. long, campanulate to cylindrical, thickened at base, indumentum as in ♂ flowers; corolla as in ♂ flowers; staminodes 4–5, filaments 1.5–2.5 mm. long, sterile anthers 0.8–1.0 mm. long; ovary c. 2 mm. long, ± ellipsoid, glabrous; style 6.5–7.5 mm. long, first forked at 2.0–2.5 mm. and with stigmatic branches 3.0–4.0 mm. long, linear, lobed. Fruit c. 15 × 10 mm., ovoid-conical, apiculate, glabrous, orange when ripe, surrounded at the base by the enlarged widely campanulate many-toothed calyx; pyrene angular-ovoid, c. 10 × 7 mm., 1-seeded.

Zambia. C: Mfuwe, Luangwa R., fr. 30.xii.1968, *Astle* 5397 (SRGH). E: Chipata Distr., Luangwa Game Res., Chilongozi Pontoon, fl. 11.x.1960, *Richards 13334 (K)*. **Zimbabwe**. N: Hurungwe Distr., Mana Pools, st. ix.1969, *Gordon* 49 (SRGH). S: Chiredzi Distr., Nyahungwe, Rundi R., Gona-re-zhou Game Reserve, fl. 12.ix.1970, *Sherry* 175 (SRGH). **Malawi**. N: Karonga, fl. xi.1887, *Scott* s.n. (K). C: Dedza Distr., Njolo Village, Mua-Golomoti Rd., fl. immat. 22.x.1968, *Salubeni* 1169 (K; SRGH). S: Chikwawa Distr., Lengwe National Park, fl. xii.1970, *Hall-Martin* 1229 (K; SRGH). **Mozambique**. N: Cabo Delgado, Nangororo, fl. 1.xi.1959, *Gomes e Sousa* 4499 (COI; K; LMA; PRE; SRGH). Z: Zambézia, s. loc., st. 28.vii.1977, *Tawse* D19 (LISC; SRGH). T: Boroma, fr. xii.1891, *Menyharth* 866 (K). MS: Gorongosa Game Reserve, 9 km. from Chitengo to great barge of Urema R., fr. immat. 8.xi.1963, *Torre & Paiva* 9111 (C; COI; LISC; LMU).

Also known from S. Somalia, Kenya and Tanzania. In riverine forests, thickets, woodlands and savanna woodlands, always near water courses; up to 600 (1000) m.

4. **Cordia mukuensis** Taton in Bull. Jard. Bot. Nat. Belg. **41**: 258 (1971); in Fl. Congo, Rwanda et Burundi, Boraginaceae: 18 (1971). Type from Zaire (Shaba).

    Ehretia guerkeana De Wild. in Ann. Mus. Congo, Bot., Sér. 4, **1**: 223 (1903) "gurkeana". —Baker & Wright in F.T.A. **4**, 2: 21 (1905) "gurkeana" non *Cordia guerkeana* Loes. (1913). Type from Zaire (Shaba).

    Cordia sp. 1 — F. White, F.F.N.R.: 364 (1962).

    Cordia senegalensis sensu Taton in Fl. Congo, Rwanda et Burundi, Boraginaceae: 17 (1971) pro parte quoad syn.

Shrub or small tree up to 5 m. high, dioecious, deciduous; trunk and branches fluted, sometimes square in transverse section and ridged at the corners; bark smooth, ± shining, grey-green to grey-blue, peeling away into small thin plate-like flakes and becoming blotched, as in guava-tree; branches ± drooping, glabrous; buds tawny-tomentose. Leaves alternate and often ± crowded at the ends; petiole 0.5–1.5 cm. long, much depressed above, pubescent when young, glabrescent; lamina 5–11(15) × 2.0–4.5 cm., lanceolate, oblanceolate or obovate to elliptic-olong or elliptic, all glabrous at maturity or sometimes with a few small hairs on the midrib, acute to subacuminate at apex, cuneate and usually symmetrical but a few times slightly asymmetrical at base, with entire margins, papery to almost leathery, with 5–7 secondary nerves on each side of the midrib, usually not raised on the lower surface. Cymes arranged in usually shortly peduncled terminal panicles 1.5–4.5(7) cm. long and 1.5–5.0 cm. wide; rhachis and branches puberulous or pubescent, usually somewhat laterally compressed. Flowers ♂ or ♀, on pedicels 0.5–2.0 mm. long. ♂ flowers with buds 4–5 mm. long, obovoid; calyx 4.0–5.5 mm. long, campanulate-infundibuliform, irregularly 3–5-toothed at apex, membranous, puberulous to subglabrous outside, minutely strigose inside; corolla white or yellowish, with tube 3–5 mm. long, narrowly infundibuliform, glabrous outside and inside save by a few hairs inside, lobes 4–5, 4.5–6.0 × 1.5 mm., oblong-lanceolate, acute, ± reflexed at anthesis; stamens 4–5, inserted on the corolla-throat, filaments 3–7 mm. long, slightly widened at base, glabrous, anthers 1.0–1.6 mm. long, with broad connective and usually emarginate at apex and cleft at base up to ⅓ or even ½; ovary vestigial. ♀ flowers with buds 4–5 mm. long, ellipsoid; calyx 5.0–5.5 mm. long and c. 3 mm. in diam., cylindrical, leathery and thickened at base with 3–5 reflexed and soon eroded teeth, pubescent to subglabrous outside, strigose inside; corolla with tube 3.0–4.5 mm. long, cylindrical, lobes 4–5, 3.5–5.5 ×

1.5–2.5 mm., acute to retuse at apex, besides as in ♂ flowers; staminodes with 1.0–1.5 mm. long filaments and sterile anthers 0.5–1.0 mm. long; ovary 1.5–2.5 mm. long, ellipsoid, glabrous; style exsert, 4–6(7) mm. long, first forked at 0.5–1.0 mm. from the base and with stigmatic branches 2.5–3.5 mm. long, linear. Fruit up to 15 × 10 mm., ovoid-conical, apiculate, glabrous, surrounded at base by the enlarged widely campanulate many-toothed calyx up to 5 mm. long and 8 mm. wide at the mouth; pyrene ovoid-angulous, c. 10 × 7 mm., with 1 or rarely 2 seeds.

Zambia. B: Kaoma (Mankoya), near Kaoma (Mankoya) Boma, fr. 23.ii.1952, *F. White* 2125 (FHO; K). N: Luangwa Valley, st. 27.vi.1968, *Astle* 5341 (SRGH). W: Luanshya, fl. 29.xi.1955, *Fanshawe* 2632 (K; LISC). C: Kabwe (Broken Hill), fl. 11.xi.1963, *Fanshawe* 8105 (FHO; K). S: Choma Golf Course, fr. 3.ii.1963, *Astle* 2023 (SRGH). **Zimbabwe**. N: Hurungwe (Urungwe), Magunge, fl. 15.xi.1952, *Lovemore* 294 (K; LISC; SRGH). **Malawi**. S: Mulanje Distr., Nantapu village, fr. 18.v.1964, *Salubeni* 316 (K; SRGH). **Mozambique**. T: Mágoè, R. Metendezi, 55 km. from Mágoè on the Rd. to Zumbo, fr. 3.iii.1970, *Torre & Correia* 18171 (COI; LISC; LMA; LMU; WAG).

Also recorded from Zaire. In riverine thicket and open woodland usually on termite mounds; 300–1000 m.

5. **Cordia myxa** L., Sp. Pl.: 190 (1753). —Hutch. in Kew Bull. **1918**: 219, t. on p. 220 fig. 1 (1918).
—Taton in Fl. Congo, Rwanda et Burundi, Boraginaceae: 16 (1971). Type based on a plant that was cultivated in Uppsala, probably from Egypt.

Small tree up to 10 m. high with spreading crown and long glabrous or subglabrous young shoots; leaf scars very prominent. Leaves alternate; petiole 1.5–4.5 cm. long, stout, glabrous or subglabrous; lamina 5–16 × 5–16 cm., broadly ovate to circular, glabrous or hairy only on the nerves above, pubescent to subglabrous but hairy at the axils below, rounded to shortly and obtusely acuminate at apex, obtuse to subcordate, rarely subacute at base, margins entire to slightly dentate or crenate-dentate, leathery, with 3–7 nerves from the base, the middle one with 3–5 secondary nerves on each side, net-veined. Cymes arranged in few-flowered lax terminal panicles, 3–8 cm. long, sometimes on short lateral branches; rhachis and branches glabrous or subglabrous. Flowers variable, apparently all ♂ but probably functionally unisexual and bisexual; buds 4–5 mm. long, obovoid, puberulous or pubescent on the upper part and sometimes brush-like at apex. Calyx 5–8 mm. long, narrowly campanulate, irregularly 3–5-toothed, glabrous or subglabrous outside, densely pubescent inside. Corolla cream-coloured, glabrous; tube 3.5–6.5 mm. long, cylindrical; lobes 4–6, 5–7 × 2.0–3.5 mm., elliptic to obovate, obtuse at apex, reflexed and ± coiled. Stamens inserted at the corolla-throat; filaments 1.5–3.5 mm. long, hairy at the base; anthers 1.8–2.0 mm. long, oblong-sagittate, sometimes sterile. Ovary c. 3 mm. long, ellipsoid; style c. 8 mm. long in total length with stigmatic branches c. 5 mm. long, oblong, wide flattened with ± deeply toothed or lobed margins. Fruit up to 30 mm. long, ovoid or globose, apiculate, glabrous, yellowish when ripe, surrounded at base by the widely enlarged campanulate calyx; pyrene c. 25 × 12 × 8 mm., more or less subwinged, 1–2-seeded.

Zimbabwe. E: Mutare, west suburbs, fr. 28.vii.1962, *Chase* 7922 (K; LISC). **Malawi**. S: Limbe, st. i.1955, *Lewis* 147 (FHO). **Mozambique**. N: Angoche (António Enes), city precincts, fr. 22.x.1965, *Mogg* 32348 (LISC; SRGH). Z: Quelimane, near docks, fl. & fr. 14.x.1965, *Mogg* 32394 (LISC). T: Báuè, fl. immat., 6.ix.1964, *Gomes e Sousa* 4826 (COI; K; PRE). MS: 3 km. from Bandula to Macequece, fr. 7.i.1948, *Mendonça* 3631 (LISC). GI: Chibuto, Baixo Changane, fr. 26.viii.1963, *Macêdo & Macúacua* 1150 (K; LMA).

Introduced and cultivated several centuries ago in several warm regions of the world and at the present apparently naturalized in some of them. In the Flora Zambesiaca area sporadically cultivated in Zambia, Zimbabwe and Malawi and very frequently cultivated and apparently subspontaneous in Mozambique. In woodland and savanna woodland, mainly on alluvial soil; up to 1480 m.

6. **Cordia monoica** Roxb., Pl. Coromandel **1**: 43, t. 58 (1795). —DC., Prod. **9**: 479 (1845). Type: *Roxburgh* drawing no. 200 at Kew.
Cordia ovalis Hochst. ex A.DC., Prodr. **9**: 479 (1845). —Baker & Wright in F.T.A. **4**,2: 15 (1905).
—Brenan & Greenway, T.T.C.L.: 76 (1949). —Rattray & Wild in Rhod. Agric. Journ. **52**: 495 (1955). —Gomes e Sousa, Dendrol. Moçamb. Estudo Geral **2**: 657, t. 212 (1967). —Taton in Fl. Congo, Rwanda et Burundi, Boraginaceae: 13 (1971). —Palmer & Pitman, Trees of Southern Afr. **3**: 1939, cum fig. & cum photogr. (1973) pro parte excl. syn. —R.B. Drumm. in Kirkia **10**: 271 (1975). —Compton, Fl. Swaziland: 479 (1976). —Palgrave, Trees of Southern Afr.: 801, cum tab. (1977). Type from Ethiopia.

Cordia quarensis Gürke in Engl., Pflanzenw. Ost-Afr. **C**: 335 (1895); in Bot. Jahrb. **28**: 308 (1900). —Baker & Wright in F.T.A. **4**, 2: 17 (1905). —Brenan & Greenway, T.T.C.L.: 76 (1949). Type from Tanzania.

Shrub or small tree up to 5(7) m. high; branchlets angular, usually rough and minutely fulvous-tomentose when young with stellate, ± branched or simple hairs, glabrescent. Leaves alternate; petiole 0.5–2.5(3.5) cm. long, flat or slightly canaliculate above; lamina 3–10(14) × 2–8(10) cm., ovate, obovate or subcircular, sometimes elliptic, rough and sometimes also with some stellate or variously branched hairs above, densely pubescent to rough and often with stellate or branched hairs mainly on the nerves below, obtuse or rounded and sometimes shortly acuminate at apex, obtuse to truncate, rarely acute at base, with subentire, crenate to crenate-serrate margins, leathery and with 3–5 secondary nerves on each side of the midrib, the lower ones basal or nearly so and stronger than the other. Cymes arranged in usually few-flowered terminal or very rarely axillary panicles 1.5–5(7) cm. long; rhachis and branches minutely fulvous-tomentose. Flowers polygamous on pedicels up to 2 mm. long. Calyx 6–8 mm. long, narrowly campanulate or infundibuliform, irregularly 3–5-toothed, rough to minutely fulvous-tomentose outside, sparsely pubescent to glabrous inside. Corolla yellow, glabrous; tube 4–7 mm. long, cylindrical; lobes (4)5(6), 4–6(7) × 1–2 mm., oblong to narrowly obovate, rounded at apex, reflexed. Stamens (or staminodes) as many as the corolla lobes, inserted at the corolla-throat; filaments 1.5–4.0 mm. long, slender, glabrous or with a few long hairs at the base; anthers 1.5–2.0 mm. long on the ♂ flowers and c. 1 mm. long on the ♀ and ♂ ones, but on the latter ones often almost sterile. Ovary c. 2 mm. long, ovoid-conical, glabrous, on the ♂ flowers vestigial; style 7–9 mm. in total length, first-forked at 1.5–3.5 mm. and with stigmatic branches 2.5–5.0 mm. long, linear. Fruit 15–22 × 10–15 mm., ovoid or ellipsoid, apiculate, glabrous, orange when ripe, surrounded at the lower third by the widely enlarged campanulate calyx; mesocarp fleshy; pyrene c. 10 × 7 mm., usually deeply cleft at the apex, ± circular in cut across, 1–3-seeded.

Zimbabwe. N: between Glendale and Bindura, Hillymead Farm, fl. 14.xii.1978, *Arkell* 1 (SRGH). E: Chimanimani Distr., between Birchenough Bridge and Piriviri (Biriwiri), Chikwizi R., fl. & fr. 15.xii.1963, *Chase* 8091 (K; LISC). S: Gwanda Distr., Doddieburn Ranch, st. 5.v.1972, *Pope* 639 (LISC; SRGH). **Mozambique**. N: between Pemba (Porto Amélia) and the Maringanha lighthouse, fl. & fr. 21.iii.1960, *Gomes e Sousa* 4550 (COI; K; SRGH). T: 3 km. from Tete to Changara, fr. 13.ii.1968, *Torre & Correia* 17532 (LISC; LMA; MO; SRGH). GI: 8 km. from Massingir to Macuma, fl. 14.xi.1970, *Correia* 1965 (LMU). M: Changalane, R. Umbeluzi, fr. 1.vi.1948, *Torre* 7925 (LISC; LMU).

Also recorded from Arabia, tropical east Africa, east of Zaire, Burundi, Angola, S. Africa, Swaziland, Mauritius and tropical Asia. Usually on sandy or alluvial soils near rivers, sometimes on anthills; up to 1200 m.

Palgrave (1977) records this species for Botswana (N) at Maun region but I did not see any specimens from that area.

7. **Cordia stuhlmannii** Gürke in Engl., Pflanzenw. Ost-Afr. **C**: 335 (1895); in Bot. Jahrb. **28**: 308 (1900). —Baker & Wright in F.T.A. **4**, 2: 16 (1905). Type: Mozambique, Quelimane (Quilimane), *Stuhlmann* Coll. I 399 (B†, holotype).

A shrub or small tree up to 8 m. high, probably dioecious; branches angular, slender, pubescent, cream or grey, becoming brown or greyish-brown and lenticellate with age, with prominent leaf-scars. Leaves alternate; petiole 0.5–2.5 cm. long, slender, deeply and narrowly canaliculate, appressed hispid; lamina 3–12 × 1.5–4.5 cm., lanceolate to oblanceolate with scattered ± appressed scarcely bulbous-based hairs 0.5–1.5 mm. long above, more densely below, acute to acuminate at apex, cuneate to obtuse at base, with margins denticulate or crenate, each tooth at the end of a secondary or tertiary nerve with mucro 0.3–1.5 mm. long, papery, blackish-green above, olivaceous below, with 3–6 secondary nerves on each side of the midrib, the lower ones opposite and subbasal. Inflorescence a panicle 5–11 cm. long, slender, lax, long pedunculate, terminal but apparently axillary, with peduncle, rhachis and branches shortly hispid. Flowers ♂ & ♀, on pedicels up to 2 mm. long. Calyx (5)6–7 mm. long, cylindrical or trumpet-shaped, irregularly 3–5-toothed, with short, ± dense appressed hairs outside, densely clothed inside. Corolla yellowish, glabrous outside, densely hairy at the throat inside; tube as long as the calyx, cylindrical; lobes 4, 4–5 × 1.7 mm., oblong-lanceolate, obtuse. Stamens (and staminodes in ♀ flowers) 4, inserted at c. 1 mm. from the top of the corolla tube; anthers 1.7–1.8 mm. long, oblong (in ♀ flowers 0.8 mm. long, sterile); filaments c. 5 mm. long, very slender, glabrous (in ♀ flowers c. 2 mm. long). Ovary c. 1.5 mm. long, cylindric-conical, glabrous (in ♂ flowers vestigial); style 7–8 mm. long, first-forked at c. 2.5 mm. and with stigmatic branches c. 2.5 mm. long, clavate. Fruit not seen.

Mozambique. Z: 23 km. from Namacurra to Maganja, fl. 27.i.1966, *Torre & Correia* 14174 (COI; K; LISC; LMU; WAG). MS: Bícuzi, Reserva Florestal do Mucheve, fl. 28.xi.1967, *M.F. Carvalho* 987 (LISC).
Known only from Mozambique. In thicket on sandy soils, sometimes on termitaries; up to c. 200 m.

8. **Cordia mandimbana** E. Martins in Garçia de Orta, Sér. Bot. **9**, 1–2: 71, t. 1 (1988). Type: Mozambique, Mandimba, Ngami Valley, *Hornby* 2401 (PRE, holotype).

Large tree probably dioecious, deciduous; branches minutely and densely pubescent, glabrescent. Leaves alternate; petiole 0.5–4.0 cm. long, covered with short, sharp, ± spreading rigid hairs; lamina 4.5–7.5 × 3–4 cm. on fertile branches, up to 15 × 8 cm. on strong sterile shoots, ovate to obovate, rough to shortly appressed hispid above, ± densely pubescent below, dark brown above, grey-brownish below when dry, acute to rounded and apiculate or abruptly acuminate at apex, obtuse to rounded at base, margins ± undulate and remotely denticulate, papery to leathery, with 4–6 secondary nerves on each side of the midrib, flat or slightly deepened and clothed with short fulvous bristles above, prominent and pubescent below. Cymes arranged in a panicle 5–8 cm. long, slender, lax, long-pedunculate, with divaricate branches, minutely hispid, terminal. Flowers subsessile. ♀ flowers with calyx 3.5–5.0 mm. long, narrowly campanulate, irregularly 3–5-dentate, minutely strigose outside, silky inside, hardly leathery; corolla (? whitish) glabrous outside, with some scattered long hairs inside on the upper half of the tube and base of the lobes, tube 4.5–5.5 mm. long, cylindrical, lobes 5, 4.0–4.5 × 1.5–2.0 mm., oblanceolate to obovate, obtuse to rounded at apex, narrowed at base, arching and reflexed at anthesis; staminodes 5, inserted near the apex of the corolla tube, filaments 3.0–3.5 mm. long, glabrous; ovary roughish; style 7–8 mm. long, first-forked at 2–3 mm. and with stigmatic branches 3–5 mm. long, linear, flattened and denticulate. ♂ flowers and fruits not seen.

Mozambique. N: 3 km. N. of Mandimba, Ngami Valley, fl. 19.xi.1941, *Hornby* 2401 (LMA: PRE). Only known from the type material. In open woodland near streams.

9. **Cordia pilosissima** Baker in Kew Bull. **1894**: 28 (1894). —Hiern, Cat. Afr. Pl. Welw. **1**: 713 (1898). —Baker & Wright in F.T.A. **4**, 2: 18 (1905). —F. White, F.F.N.R.: 364 (1962). —Fanshawe in Kirkia **6**: 177 & 178 (1968). —Taton in Fl. Congo, Rwanda et Burundi, Boraginaceae: 14 (1971). —Palmer & Pitman, Trees of Southern Afr. **3**: 1941 (1973). —R.B. Drumm. in Kirkia **10**: 271 (1975). —Palgrave, Trees of Southern Afr.: 801, cum fig. (1977). —Hall-Martin & Drumm. in Kirkia **12**: 177 (1980). Type from Angola.
    Cordia kirkii Baker, loc. cit. (1894). —Gürke in Engl., Pflanzenw. Ost-Afr. **C**: 335 (1895). —Schinz in Denkschr. Akad. Wiss. Wien, Math.-Nat. Kl. **78**: 436 (1905). Type: Mozambique, "Tette", *Kirk* s.n. (K, holotype).

Multiple-stemmed shrub, sometimes scrambling, or small tree up to 10(15) m. high, dioecious, deciduous; branches angular-ribbed and densely hairy when young, glabrescent and with a little prominent leaf-scars. Leaves alternate; petiole 1–4(6) cm. long, usually tomentose-villous; lamina 6–18(23) × 5–15(20) cm., ovate or widely ovate to circular, pubescent or hairy but velutinous when young and sometimes roughish at maturity above, hairy or tomentose below, subacute to rounded and sometimes retuse at apex, truncate to subcordate, rarely obtuse at base, with ± crenate-denticulate and sometimes repand margins, papery, with 4–7 pairs of secondary nerves, sometimes 3–5-nerved from the base, venation slightly impressed above, prominent below. Cymes arranged in usually subsessile, short, ± contracted terminal panicles; rhachis and branches densely villous. Flowers ♂ or ♀, on pedicels 0.5–3.0 mm. long, villous. ♂ flowers with calyx 6–10 mm. long, infundibuliform, prominently nerved, hairy or villous outside, pubescent inside, irregularly dentate; corolla white to yellow, glabrous, tube as long as the calyx, cylindrical, limb c. 15 mm. in diam. with 5 lobes 5–8(10) × 2–3(4) mm., narrowly obovate, crisped, truncate or sometimes bilobed at apex; stamens 5, inserted at the mouth of the corolla, filaments 4.0–5.0 mm. long, slender, villous to the base, anthers 2.0–2.5 mm. long, oblong; ovary vestigial, without style. ♀ flowers similar to the ♂ ones but with filaments c. 2.5 mm. long and sterile anthers c. 1.5 mm. long; ovary c. 1.5 mm. long, cylindric-conical, glabrous; style c. 10 mm. long, first branched at c. 4 mm. and with stigmatic branches 3–6 mm. long, linear, flattened, papillose. Fruit up to 27 × 16 mm. when dry, ovoid or elliopsoid, apiculate, glabrous; fruiting calyx enlarged up to 20 mm. long and 25 mm. in diam. at the mouth, with the margin folded and revolute, eroded and usually fallen at maturity, slightly ribbed; mesocarp fibrous; pyrene c. 15 mm. long, ± circular in cross section, 1-seeded.

Caprivi Strip. Mpilila Is., st. 13.i.1959, *Killick & Leistner* 3359 (K). Botswana. N: Chobe Nat. Park, Kasane, st. 1.ix.1970, *Mavi* 1163 (K; SRGH). Zambia. C: Lusaka Distr., between Kafue and Kafue R., fl. 23.xi.1959, *Drummond & Cookson* 6761 (K; LISC; SRGH). S: Livingstone Distr., Victoria Falls, Eastern Cataract, fr. 5.ii.1963, *Bainbridge* 720 (FHO; SRGH). Zimbabwe. N: Mazowe Distr., Shamva, Chipoli Farm, fl. 25.xi.1958, *Moubray* 71 (K; SRGH). W: Hwange Distr., Lutope-Gwai R. Junction, fr. immat. 26.ii.1963, *Wild* 6015 (K; LISC; SRGH). C: Kudu River Ranch, Munyati (Umniati) R., st. 28.iii.1970, *Burrows* 435 (SRGH). Malawi. S: Chikwawa Distr., Lengwe Nat. Park, st. 16.viii.1970, *Hall-Martin* 903 (FHO; SRGH). Mozambique. T: Cahobra (Cahora) Bassa, Chetima (Estima), near R. Sanangoé, fr. immat. 19.xi.1973, *Macêdo* 5385 (LISC; LMU). MS: Gorongosa Nat. Park, Rd. 5, fr. immat. i.1972, *Tinley* 2357 (K; LISC; SRGH).

Also in Angola and Zaire (Haut-Katanga). In thicket and open woodland on river banks and in riparian deciduous thicket in ravines near rivers; 100–975 m.

10. **Cordia grandicalyx** Oberm. in Ann. Transv. Mus. **17**: 219, cum tab. (1937). —Codd, Trees & Shrubs Kruger Nat. Park: 161, t. 148 (1951). —Palmer & Pitman, Trees of Southern Afr. **3**: 1941, cum fig. (1973). —R.B. Drumm. in Kirkia **10**: 271 (1975). —Palgrave, Trees of Southern Afr.: 800, cum fig. (1977). Type from S. Africa (Transvaal).

Multiple-stemmed shrub or small tree up to 7 m. high, dioecious, deciduous; branches tomentose, glabrescent and with prominent leaf-scars. Leaves alternate; petiole 0.5–3.0(5.0) cm. long, subterete, tawny-pubescent or tawny-hairy; lamina 4–13(16) × 3.5–10.0(12.5) cm. very variable in shape, widely ovate or widely obovate to reniform, roughish but glandular-pubescent when young above, tomentose below, usually rounded at apex, obtuse to subcordate at base, margins entire, slightly crenate to minutely dentate, leathery, with 3–4 secondary nerves on each side of the midrib, the basal ones reaching the margins on the upper half, or 3–5-nerved from base, very prominently net-veined below. Cymes arranged in short ± contracted terminal panicles; rhachis and branches minutely tawny-tomentose; pedicels 2–11 mm. long. ♂ flowers with calyx 13–14 mm. long, tubular-infundibuliform, prominently nerved, minutely tawny-tomentose outside, sparsely pubescent inside, with 3–6 triangular teeth 1.5–2.5 mm. long; corolla sordid with tube 15–17 mm. long, cylindrical, glandular-puberulous outside, glabrous inside save by some whitish long hairs below the insertion of stamens, limb 15–20 mm. in diam. with (4)5(6) oblanceolate, outspread lobes; stamens inserted near the apex of the corolla tube, filaments 6–8 mm. long, slender, hairy to the base, anthers 2.5–4.0 mm. long, oblong; ovary vestigial, without style. ♀ flowers with calyx 9–11 mm. long, besides as in ♂ flowers; corolla with tube c. 12 mm. long, besides as in ♂ flowers; staminodes with filaments 2.5–3.5 mm. long and sterile anthers 2.0 mm. long; ovary c. 1.5 mm. long, ovoid, glabrous; style 18–22 mm. long, first branched at c. 11 mm. from the base and with stigmatic branches 6–10 mm. long, linear-filiform, glabrous. Fruit c. 40 × 35 mm., ovoid, with a horn-shaped point c. 15 mm. long, glabrous; fruiting calyx widely campanulate, enlarged up to 35 mm. wide at the mouth, with crisp and revolute margin, pubescent or tomentellous outside, very prominently nerved outside mainly on the upper half; mesocarp fibrous; pyrene coarsely trapezoidal in cross section, 1-seeded.

Zimbabwe. W: Matopos, Whitewaters, fl. 4.x.1952, *Plowes* 1485 (K; LISC; PRE; SRGH). E: Chimanimani Distr., Umvumvumvu Irrigation Scheme, Hot Springs Rd., fr. immat. 16.xi.1952, *Chase* 4698 (K; LISC; PRE; SRGH). S: Doddieburn Ranch, fr. 27.iv.1924, *Davison* 35 (FHO; PRE). Mozambique. N: Nampula Prov., between Lalaua and Muitetere, R. M'tar, fl. 17.x.1967, *Macêdo* 2761 (LMA).

Also in S. Africa (Transvaal). On river banks and on rocky soils; 350–1460 m.

11. **Cordia torrei** E. Martins in Garçia de Orta, Sér. Bot. **9**, 1–2: 71, t. 2 (1988). Type: Mozambique, Monapo, 7 km. from Namialo to Meserepane, *Torre & Paiva* 9261 (LISC, holotype).

Shrub c. 5 m. high, dioecious, deciduous; branches densely clothed with grey-brownish appressed 2-branched, medifixed and shortly pediculated trichomes 0.5–1.0 mm. long, soon glabrescent, mature light grey and marked with prominent leaf-scars. Leaves alternate; petiole 1.5–3.3 cm. long, slender, with indumentum as on the young branches; lamina 5.5–12.0 × 3.0–7.5 cm., obovate to elliptic, on mature leaves glabrous or subglabrous above, shortly woolly below, the trichomes easily removable, green-brownish or brown above, grey-yellowish or grey-brownish below in sicco, subacute, acute or acuminate to obtuse acuminate or rounded at apex, subacute to rounded at base, margins subentire to irregularly serrate-dentate, papery, with 4–6 secondary nerves on each side of the midrib, the lower ones subopposite appearing 3–5-nerved from or near base. Cymes arranged in a panicle 3–9 cm. long and 4–8 cm. wide with indumentum like the branches, peduncle as long as half of the whole, slender. Flowers subsessile. ♀ flowers with calyx 7–9 mm. long, narrowly infundibuliform, slightly ridged, minutely tomentose-woolly outside, puberulous inside, 4–5-toothed, the teeth soon eroded, papery; corolla (? white) glabrous, tube 6.5–8.0 mm. long, 1.5 mm. wide, cylindrical, lobes 4, 6.0–6.5 × 2.5–3.0 mm., oblanceolate, obtuse; staminodes 4, filaments c. 1 mm. long, sterile anthers c. 0.7 mm. long; ovary c. 2 mm. long, ovoid; style 10–11 mm. long, first-forked at 3–4 mm. and with stigmatic branches 4–6 mm. long, linear but a little flattened, denticulate. ♂ flowers and fruits unknown.

Mozambique. N: Nampula (Moçambique), Monapo, 7 km. from Namialo on the track to Meserepane, fl. 24.xi.1963, *Torre & Paiva* 9261 (C; COI; K; LISC; LMU).
Also in Tanzania. In woodland with *Sterculia appendiculata, Acacia nigrescens* and *Pterocarpus angolensis*, c. 180 m.

12. **Cordia sinensis** Lam., Tab. Encycl. Méth. Bot. **1**: 423 (1792). —Poir., Encycl. Méth. Bot. **7**: 49 (1806). —Heine in Adansonia, Sér. 2, **8**: 186 (1968). — R.B. Drumm. in Kirkia **10**: 271 (1975). —Palgrave, Trees of Southern Afr.: 802, cum tab. (1977). TAB.**21**. Type from India.
 Cordia rothii Roem. & Schult. in Syst. Veget. **4**: 798 (1819). —DC., Prodr. **9**: 480 (1845). —Baker & Wright in F.T.A. **4**, 2: 18 (1905). —Schinz in Denkschr. Akad. Wiss. Wien, Math.-Nat. Kl. **78**: 436 (1905). Type from India.
 Cordia quercifolia Klotzsch in Peters, Reise Mossamb. Bot. **1**: 247, t. 43 (1861). Type: Mozambique, Tete, *Peters* s.n. (B†, holotype).
 Cordia gharaf Ehrenb. ex Aschers. in Sitz.-Ber. Ges. Nat. Fr. Berl. **1879**: 46 (1879); in Verh. Bot. Ver. Prov. Brand. **21**: 69 (1880). —Gürke in Engl., Pflanzenw. Ost-Afr. **C**: 335 (1895). —Brenan & Greenway, T.T.C.L.: 76 (1949). —I.M. Johnston in Journ. Arnold Arb. **37**: 292 (1956). — Friedrich-Holzhammer in Merxm., Prodr. Fl. SW. Afr. **119**: 2 (1967). Type from Yemen.
 Cordia ovalis sensu Palmer & Pitman, Trees of Southern Afr. **3**: 1939 (1973) pro parte quoad syn.

A shrub usually up to 4 m. or a bushy tree up to 8 m. high; branchlets angular, fulvous-tomentellous to subglabrous when young, glabrescent, often cream becoming grey or brown and lenticellate with age. Leaves opposite, subopposite or alternate; petiole 0.3–1.0(1.5) cm. long usually with some long pale hairs; lamina 3–9 × 1.0–4.5 cm., narrowly obovate to elliptic, rarely oblanceolate, with minute bulbous-based appressed or slightly ascending hairs and sometimes with scattered white bristles to subglabrous but often dotted with groups of cystoliths above, with minute bulbous-based hairs appressed in all directions or pubescent to subglabrous but usually with dense long hairs on the axils of the secondary nerves and along the midrib (sometimes not much evident) below, obtuse to rounded and sometimes retuse at apex, acute to obtuse and often slightly asymmetrical at base, margins entire to crenate or serrate to the upper half, papery to leathery, greyish-green or yellowish-green, with 5–7 secondary nerves on each side of the midrib. Panicle 1.5–6.0 cm. long, little-branched, terminal or axillary; rhachis and branches minutely fulvous-tomentose with scattered long pale hairs to pubescent, often glabrescent. Flowers ♂ on pedicels 0.5–1.5 mm. long. Calyx 4–5 mm. long, narrowly campanulate, irregularly 3–5-toothed, pubescent or minutely fulvous-tomentose outside, strigose inside. Corolla white, glabrous; tube 2.5–4.0 mm. long, cylindrical; lobes 4, 3.0–4.0 × 1.2–2.0 mm., oblong to narrowly obovate, rounded at apex, reflexed. Stamens 4, inserted at the corolla-throat; filaments 2–3 mm. long, glabrous; anthers c. 1.5 mm. long, oblong. Ovary c. 2 mm. long, ovoid, glabrous; style 5–7 mm. long, first-forked at 2.0–2.5 mm. and with stigmatic branches 2–3 mm. long, clavate. Fruit 12–15 × 8–11 mm., ovoid or ellipsoid, apiculate, glabrous, yellow, orange or reddish when ripe, surrounded at the lower third by the widely enlarged, usually cream, campanulate calyx; mesocarp fleshy; pyrene c. 8 × 7 mm., ± rhomboidal or quadrangular in cross section, 1–3 seeded.

Caprivi Strip. Mpilila I., fr. 13.i.1959, *Killick & Leistner* 3352 (K; SRGH). **Botswana**. N: Thamalakane R., fr. 12.ii.1972, *Smith* 183 (K; LISC; SRGH). SE: Orapa, "Lake" David Grey, st. 25.v.1976, *Allen* 396 (PRE). **Zambia**. C: Lusaka Distr., 11 km. N. of Homestead Lochinvar Ranch, fr. 24.iv.1962, *Angus* 3146 (LISC; SRGH). S: 112 km. upstream from Kariba Gorge, fr. v.1957, *Scudder* 79 (K; SRGH). **Zimbabwe**. N: Gokwe Distr., Zambezi Valley, fr. vi.1956, *Davies* 1980 (K; SRGH). W: Shangani Reserve, *Davies* in GHS 32066 (SRGH). C: Manyame (Hunyani) R., 19 km. from

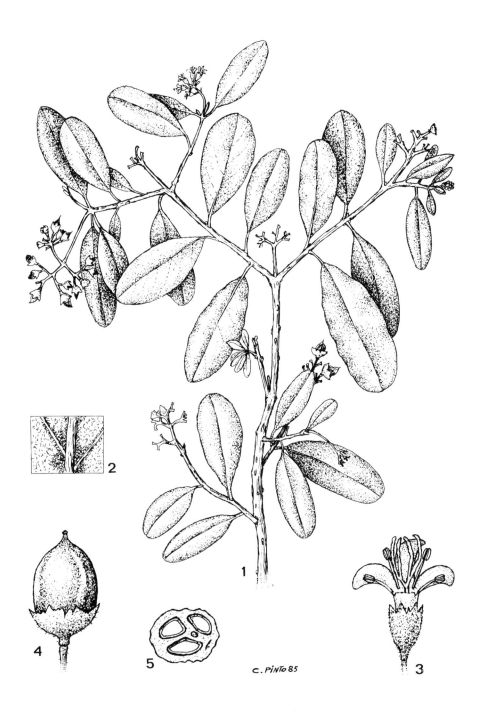

Tab. 21. CORDIA SINENSIS. 1, flowering/fruiting twig (×$\frac{1}{2}$); 2, portion of inferior leaf surface to show indumentum (× 3); 3, flower with corolla lobe removed (× 3); 4, fruit (× 2); 5, pyrene in cross section (× 2), all from *Rosa* 101.

Harare, fr. 26.iii.1933, *Pardy* 13/33 (FHO; SRGH). E: Chimanimani, Hot Springs, fr. 3.v.1948, *Chase* 708 (BM; K; SRGH). S: Great Zimbabwe Nat. Park, fr. 16.iii.1971, *Chiparawasha* 370 (K). **Mozambique**. N: Cabo Delgado, Palma, fr. 27.ix.1948, *Pedro & Pedŕogao* 5420 (LMA). T: Zumbo, Panhame, fl. 15.i.1950 et fr. 15.iii.1950, *Melo* 14 (LMA); Tete, fr. 18.iii.1966, *Torre & Correia* 15210 (C; LISC; LMU; PRE). MS: Chemba, Tambara, near Fort, fr. 15.v.1971, *Torre & Correia* 18459 (COI; K; LISC; LMA). M: 3.5 km. from Goba to Goba Fronteira, fl. & fr. 15.v.1975, *Marques* 2768 (K; LISC; LISU; LMU; SRGH).

Widespread in the drier parts of tropical Africa, from Senegal to Egypt and Sudan and thence through E. Africa to the Flora Zambesiaca area and Angola and Namibia and, probably, to Swaziland and S. Africa (Transvaal and Natal); also in tropical Asia from Arabia to India. In grasslands, open deciduous woodlands and riverbanks, usually on sandy or rocky soils or on termitaries; up to c. 100 m. Sometimes cultivated; the fruits are edible and mainly consumed by the Asian community.

The cultivated specimens are probably the only typical ones from the Flora Zambesiaca area and they show a great homogeneity of features probably due to their direct introduction from India. That is not the case with the indigenous specimens which show a great variability even on the same Herbarium sheet. While the floral characters as well as those of the pyrene are more or less constant, the vegetative characters are not so. The specimens from the Zambezi Valley (Lake Kariba region to Mutarara on Lower Zambezi, particularly numerous in the Tete-Boroma region), exhibit characters that agree with *C. quercifolia*, as do most of the specimens from Chimanimani (Melsetter) and Chipinge (Chipinga) districts in Zimbabwe (E). The specimens from Caprivi, Botswana (N & SE), Zambia (S) and Zimbabwe (N, W, C & S) exhibit characters that suggest *C. monoica*. The latter grow very often, but not always, on termite mounds, in *C. quercifolia* this is less often observed.

In spite of both the diversity and lack of more comprehensive studies, I prefer to include all of them provisionally in the same taxon.

13. **Cordia caffra** Sond. in Linnaea **23**: 81 (1850). —C.H. Wright in F.C. **4**, 2: 4 (1904). —Sim, For. Fl. Port. E. Afr.: 90 (1909). —Wood, Natal Plants **6**, 1: t. 508 (1909). — J.H. Ross, Fl. Natal: 297 (1972). —Palmer & Pitman, Trees of Southern Afr. **3**: 1937 cum tab. & photogr. (1973). —Palgrave, Trees of Southern Afr.: 799 cum tab. (1977). Syntypes from S. Africa. (Cape Prov. and Natal).

Shrub or small tree up to 7 m. high, deciduous; branches greyish, glabrous; branchlets rusty-pubescent but soon glabrescent. Leaves alternate; petiole 2–5 cm. long, slender, canaliculate, glabrous; lamina 3.0–9.5 × 2.5–6.0 cm., lanceolate to widely ovate or elliptical, minutely rusty-tomentose on the very young leaves but glabrous on the mature ones, slightly discolorous, acute to acuminate, rarely obtuse at apex, obtuse to truncate, rarely acute but often asymmetrical at base, with crenate-serrate margins, papery, with 4–6 secondary nerves on each side of the midrib, not prominent on lower surface. Cymes arranged in a small lax panicle, terminal and often on short lateral branches; rhachis and branches at first pubescent, becoming glabrous. Flowers ♀ and ♂ on pedicels 1–5 mm. long; buds globose, at last obovoid. Calyx 4–5 mm. long, tubular-campanulate, irregularly 3–5-toothed, membranous, minutely and sparsely pubescent outside, strigose inside mainly in the upper part. Corolla whitish, yellowish or greenish; tube 2.5–3.5 mm. long, campanulate or tubular-campanulate, glabrous outside and inside save by some whitish hairs below the insertion of the stamens; lobes 4–5(6), 4.5–6.0 × 2.0–3.5 mm., lanceolate to oblanceolate, acute to obtuse at apex with the margins sinuate, glabrous, reflexed. Stamens 4–5(6), inserted at apex of the corolla tube; anthers 1.3–1.7(2.4) mm. long, oblong or ovate; filaments in ♂ flowers very variable in length, even in the same panicle, usually 1.5–4.5 mm. long, in ♂ flowers 1.8–2.2 mm. long, glabrous or slightly hairy at base. Ovary in ♂ flowers c. 1.5 mm. long, globose to conical, in ♀ flowers 2.0–2.5 mm. long, ovoid to obovoid, acute, glabrous; style in ♂ flowers lacking, in ♀ flowers 5.0–6.5 mm. long, first-forked at 0.5–1.5 mm. and with stigmatic branches c. 3 mm. long, linear-flattened. Fruit c. 14 × 9 mm., ovoid, apiculate, glabrous, orange or red when ripe, surrounded at base by the widened cup-shaped calyx; pyrene c. 12 × 8 × 5 mm., laterally compressed, 1–2-seeded.

Mozambique. GI: Manjacaze, Chidinguel, near lighthouse, fl. & fr. 11.xii.1944, *Mendonça* 3388 (C; K; LISC; LMU). M: Namaacha, near Goba, fl. 23.viii.1944, *Mendonça* 1844 (COI; LISC; SRGH); Inhaca Isl., fl. vii.1959, *Mogg* 30802 (K; SRGH).

Also in Swaziland and the eastern coastal region of S. Africa, and inland in eastern and northern Transvaal. In dune bush and coastal forest, and inland in woodland and on forest margins.

14. **Cordia sp. A**

Tree c. 10 m. high; young branches ± shortly hispid, ferruginous at the ends, the oldest ones grey, densely lenticellate. Leaves alternate; petiole 1.0–2.5 cm. long, slender, roughish to shortly hispid, canaliculate; lamina 7–12 × 2.5–4.5 cm., elliptical to oblanceolate or obovate, roughish to shortly hispid both above and below, green-blackish in dry, subacute to acuminate at apex, cuneate at base, the margins coarsely and irregularly serrate-toothed mainly on the upper half, papery, with 4–6 secondary nerves on each side of the midrib, the lower ones usually less than 25° with the midrib. Cymes arranged in small dense panicles c. 1–2 cm. long. Flowers unknown. Fruit c. 10–12 mm. long, ellipsoid, apiculate; pyrene c. 8 × 6 mm. ± quadrangular in cross section; fruiting calyx c. 5 mm. long and 8 mm. in diam. at mouth, broadly campanulate with a stipe 1–2 mm. long, margin irregularly cleft, strigose inside, subglabrous outside, densely nerved but not grooved.

Malawi. C: Lilongwe Distr., Lilongwe Nature Reserve, between Capital Hill and Lingadzi R., 1050 m., fr. 27.iii.1977, *Brummitt, Seyani & Patel* 14980 (K; SRGH).
Only known from the quoted specimen. In mixed woodland of *Combretum* spp., *Albizia harveyi* and *Lonchocarpus capassa*, on ant hill.

15. **Cordia sp. B**
Cordia sp. 2—F. White, F.F.N.R.: 364 (1962).

Branches light brown, pilose to hispid, glabrescent with age. Leaves alternate; petiole 1.5–2.5 cm. long, hairy, canaliculate; lamina 7–13 × 4–7 cm., elliptic to oblong-elliptic, glabrous above, subglabrous below, dark brown (in sicco) below, darker above, subacute to acuminate at apex, obtuse at base, margins entire to ± undulate or with some teeth or small lobes mainly to the apex, rigidly papery, with 5–7 secondary nerves on each side of the midrib, flat above, prominent below. Cymes arranged in small dense panicles 2.5–3.5 cm. long with very short or subnull peduncle, pilose to hispid. Flowers unknown. Developed fruits not seen. Fruiting calyx c. 8 mm. long and 7 mm. in diam. at mouth, campanulate, margin irregularly dentate, not grooved, glabrous outside, minutely strigose or subglabrous inside.

Zambia. N: Mporokoso Distr., between Mkupa and Chiengi, fr. 12.x.1949, *Bullock* 1218 (K).
Only known from the very poor specimen quoted above.

2. BOURRERIA P. Browne

Bourreria P. Browne, Hist. Jamaica: 168 (1756) nom. cons., *Beureria* in ind.: 492 non *Beureria* Ehret (1755). —Jacq., Enum. Pl. Carib.: 14 (1760).
    "Beurreria" Jacq., Select. Stirp. Amer.: 44 (1763).

Trees or shrubs. Leaves alternate, sometimes crowded, petiolate, simple, usually entire. Inflorescences paniculate or corymbose, sometimes few-flowered, terminal but sometimes apparently lateral. Flowers ♂, actinomorphic, subsessile or pedicellate. Calyx campanulate, closed in bud, 2–5-lobed. Corolla tube usually campanulate; lobes 5, frequently rounded, spreading or reflexed. Stamens 5, inserted on the corolla tube, sometimes about the middle, included or exserted; anthers oblong on short filaments. Style terminal on the ovary, entire or very shortly bifid; stigmas 2, capitate or peltate, or 1, bilobed. Fruit drupaceous, subglobose; pyrenes 4, bilocular, 1-seeded, the outer surface with thin woody lamellas.

A genus of about 55 species mostly in Central America and West Indies. 5 species in Africa.

Leaves pubescent to woolly below. Inflorescence densely pubescent to
    tomentose   -   -   -   -   -   -   -   -   -   -   -   1. *nemoralis*
Leaves glabrous below. Inflorescence glabrous to sparsely pubescent   -   -   2. *petiolaris*

1. **Bourreria nemoralis** (Gürke) Thulin in Nordic Journ. Bot. **7**: 415 (1987). Type from Tanzania.
    Ehretia litoralis Gürke in Engl., Pflanzenw. Ost-Afr. C: 335 (1895); in Bot. Jahrb. **28**: 310 (1900). —Baker & Wright in F.T.A. **4**, 2: 21 (1905). —Brenan & Greenway, T.T.C.L.: 77 (1949) non *Bourreria litoralis* Donn. Smith (1898). Syntypes from Tanzania.
    Ehretia nemoralis Gürke in Engl., tom. cit.: 336 (1895); loc. cit. (1900). Type as for *Bourreria nemoralis*.

A shrub or sometimes a small tree up to 5(10) m. high; branches pubescent with moderately dense short hairs, sometimes with a few longer glandular hairs. Petiole 2–6(9) cm. long, slender, canaliculate, pubescent, the hairs spreading or ± appressed, sometimes with some glandular hairs as on the branches; leaf-lamina 6–10(17) × 3–7(12) cm., ovate to obovate, pubescent above, sometimes only on the nerves, or subglabrous, pubescent to almost velvety or shortly woolly below, subacute to rounded and apiculate at apex, subacute to rounded and often ± asymmetrical at base, margins entire, papery to almost leathery, with 6–8 nerves on each side of the midrib, net-veined. Flowers arranged in panicles 7–13 cm. long, lax, bracteate, terminal, with indumentum similar to the branches and petioles; bracts falling; pedicels up to 3 mm. long, joined at the apex. Calyx 3.0–3.5 mm. long, pubescent outside, glandular hairs absent; lobes triangular, acute, as long as the tube. Corolla 3.5–4.5(5) mm. long, glabrous, whitish; tube 3.0–4.0 mm. long, narrowly campanulate; lobes 1.0–1.4 × 1.2–1.8 mm., much broadly ovate, obtuse to rounded, erect to spreading. Stamens inserted about the middle of the corolla tube, included; anthers 0.8–1.2 mm. long, subsessile. Ovary c. 1 mm. long, conical; style 1.5–2.5 mm. long, very shortly bifid, included; stigmas large, ± peltate. Fruit c. 5 × 6 mm., a little depressed, usually topped by the persistent style; pyrenes with 6–8 lamellae on the outer surface.

Mozambique. N: Cabo Delgado, 7 km. S. of Pemba (Porto Amélia), fl. & fr. 12.iii.1961, *Gomes e Sousa* 4651 (COI; K; LMA; SRGH). GI: 40 km. from Mapinhane on road to Mavume, fl. & fr. immat. ii.1939, *Gomes e Sousa* 2225 (COI; K).

Also known from Kenya and Tanzania. In thickets on coastal dunes and sometimes in *Colophospermum* or *Brachystegia* woodlands, usually on sandy soils.

Species of coastal distribution occurring sometimes to the inland at low altitudes. The specimen *Barbosa & Balsinhas* 5097 (LMA), collected at Machaíla, Chicualacuala (Limpopo) Distr., c. 250 km. distance from the sea, seems to belong to this species.

The specimens from Gaza and Inhambane Provinces (GI) usually have many glandular hairs which are rare or lacking on the specimens from Cabo Delgado and Nampula Provinces (both N. for Flora Zambesiaca). I did not see any specimen from the intermediate provinces, Zambézia, Manica and Sofala. However, glandular hairs occur, sometimes very plentiful, on specimens from Kenya and Tanzania. Therefore, in spite of the geographical discontinuity, I think that there are no reasons for taking into account the specimens seen as belonging to distinct taxa.

2. **Bourreria petiolaris** (Lam.) Thulin in Nordic Journ. Bot. **7**: 414 (1987). TAB. **22**. Type from a plant "cultivé au jardin du Roi" probably from seeds with origin in Mauritius.

    Ehretia petiolaris Lam., Encycl. Méth. Bot. **1**: 527 (1785); Tab. Encycl. Méth. Bot. **1**: 425 (1792). —DC., Prodr. **9**: 504 (1845). —Baker, Fl. Maurit.: 201 (1877). —Gürke in Engl., Pflanzenw. Ost-Afr. **C**: 336 (1895). —Baker & Wright in F.T.A. **4**, 2: 21 (1905). —Brenan & Greenway, T.T.C.L.: 77 (1949). —Munday & Forbes in Journ. S. Afr. Bot. **45**: 9 (1979). Type as above.

A shrub or sometimes a small tree up to 7(10) m. high, glabrous; branches greyish or brownish, usually slender. Leaves alternate; petiole 1.5–5.5(8) cm. long, slender, deeply canaliculate; lamina 5–11(14) × 2–6(8) cm., lanceolate or ovate to oblong-obovate, acute to rounded but usually apiculate at apex, acute to obtuse and very often asymmetrical at base, with entire margins, papery to almost leathery, brown and ochre-coloured when dry, with 6–9 secondary nerves on each side of the midrib, net-veined. Flowers arranged in pedunculate, bracteate, lax, terminal but apparently lateral panicles or corymbs (2.5)4–10 cm. long, with spreading branches; peduncle, rhachis, branches and bracts (falling) sometimes with scattered short hairs; pedicels short, joined at the apex. Calyx 3–4 mm. long with triangular lobes as long as the tube, glabrous except on the margins, rarely pubescent. Corolla 4–5 mm. long, glabrous, white; tube 3.5–4.0 mm. long, broadly campanulate; lobes 1.0–1.5 × 1.4–2.0 mm., rounded, erect to spreading. Stamens inserted about the middle of the corolla tube, included; anthers c. 1 mm. long; filaments as long as the anthers. Ovary c. 1.2 mm. long, conical; style 2.0–2.5 mm. long, subentire, included; stigmas large. Fruit 4–5 mm. long, a little depressed, topped by the persistent style; pyrenes with 6–8 lamellae.

Mozambique. N: Cabo Delgado, Palma, 4 km. from cape Delgado Lighthouse to Palma, fr. 17.iv.1964, *Torre & Paiva* 12112 (BR; COI; LD; LISC; MO). GI: Santa Carolina Isl., fl. 19.x/3–5.xi.1958, *Mogg* 28786 (K; LMU; PRE; SRGH). M: Inhaca Isl., fl. 3.xi.1962, *Mogg* 30089 (K; LMA; LMU; PRE; SRGH).

Also known from Kenya, Tanzania, Madagascar and Mauritius.

Apparently confined in Flora Zambesiaca area to the littoral scrubs mainly on coraline soils, both on the islands and on the mainland. Recorded by Brenan & Greenway, loc. cit. (1949) as "a common dominant nr. L. Chala". Probably occurring also in N. Natal, since it was collected in Ponta do Ouro on the southernmost coastal part of Mozambique.

74

Tab. 22. BOURRERIA PETIOLARIS. 1, fruiting twig (× ½), from *Torre & Paiva* 12112; 2, flower (× 3); 3, flower with part of calyx, part of corolla and 3 stamens removed (× 3), 2–3 from *Mogg* 30089; 4, fruit (× 3); 5, pyrene (× 4), 4–5 from *Torre & Paiva* 12112.

3. EHRETIA P. Browne

Ehretia P. Browne, Hist. Jamaica: 168 (1756).

Trees or shrubs. Leaves alternate, usually petiolate, simple, entire, dentate, crenate or serrate. Inflorescences paniculate or corymbose, usually terminal. Flowers ♂ or polygamous, actinomorphic, subsessile or pedicellate. Calyx 5-lobed, deeply divided, lobes imbricate in bud. Corolla tube cylindrical to widely funnel-shaped; lobes 5(6), imbricate, oblong, spreading or reflexed. Stamens 5(6), inserted at the corolla-throat, exserted; anthers oblong or sagittate. Ovary 2- or 4-locular; style terminal, ± deeply divided into 2 branches, topped by capitate or umbilicate stigmas. Fruit drupaceous, subglobose; endocarp of usually 4, 1-seeded pyrenes.

A genus of about 40 species mostly in the tropical and subtropical regions of the Old World and a few in Central America and the West Indies.

1. Corolla lobes longer than the tube, reflexed; flowers white   -   -   -   -   - 2
 - Corolla lobes as long as or shorter than the tube, ± spreading; flowers blue to mauve, rarely white   -   -   -   -   -   -   -   -   -   -   -   -   -   -   - 3
2. Leaves scabrous above, pubescent below; venation reticulate, prominent below   2. *amoena*
 - Leaves glabrous or subglabrous above and below; venation not reticulate, inconspicuous below   -   -   -   -   -   -   -   -   -   -   -   - 1. *cymosa* var. *divaricata*
3. Petiole none or up to 4 mm., rarely up to 8 mm. long; limb always entire, glabrous to shortly hispid; acarodomatia present   -   -   -   -   -   -   -   -   -   - 3. *rigida*
 - Petiole more than 4 mm. long; limb entire to crenate-dentate, pubescent and usually with capitate glandular hairs; acarodomatia absent   -   -   -   -   - 4. *obtusifolia*

1. **Ehretia cymosa** Thonn. ex Schumacher in Kongel. Dansk. Vid. Selsk. Naturvid. Meth. Afh. **3**: 149 (1828). Type from West Tropical Africa.

Var. **divaricata** (Baker) Brenan in Mem. N.Y. Bot. Gard. **9**: 4 (1954). —Taton in Fl. Congo, Rwanda et Burundi, Boraginaceae: 24 (1971). —R.B. Drumm. in Kirkia **10**: 271 (1975). Type: Malawi, near Chiradzulu (Tshiradzuvru) Mt., *Kirk* s.n. (K, holotype).
    Ehretia divaricata Baker in Kew Bull. **1894**: 28 (1894). —Gürke in Engl., Pflanzenw. Ost-Afr. **C**: 336 (1895). —Baker & Wright in F.T.A. **4**,2: 26 (1905). —S. Moore in Journ. Linn. Soc., Bot. **40**: 149 (1911). —Brenan & Greenway, T.T.C.L.: 77 (1949). Type as for *E. cymosa* var. *divaricata*.
    Ehretia cymosa sensu Palgrave, Trees of Southern Afr.: 803, cum tab. (1977).

Small to medium-sized tree up to 20(25) m. high; branches glabrous or sparsely hairy. Petiole 1.0–2.7(4.5) cm. long, slender, glabrous, puberulous or hairy; leaf-lamina 7–19 × 4–10 cm., ovate to oblanceolate, often elliptic, glabrous above except on the usually puberulous or pubescent nerves, glabrous below but with some short hairs on the acarodomatia placed at the axils of secondary and main tertiary nerves, sometimes puberulous to hairy on the nerves and rarely shortly ciliate on the margins near the base, acute to acuminate at apex, cuneate to rounded, rarely subcordate and sometimes slightly asymmetrical at base, margins entire, papery to almost leathery with 6–8 nerves on each side of the midrib. Cymes arranged in panicles sometimes corymbose up to 13 × 14 cm., lax, shortly pedunculate, with branches usually divaricate, terminal and in short lateral branches; rhachis and branches puberulous or shortly pubescent. Flowers ♂, 5-merous, scented, sessile or subsessile, the terminal ones shortly pedicellate. Calyx 2.0–2.5 mm. long, glabrous or puberulous outside, divided near base, lobes narrowly lanceolate, ciliate. Corolla white; tube 2.0–2.5 mm. long, broadly obconical; lobes 2.5–3.5 mm. long, oblong, rounded at apex, ciliate to the upper part, with slightly revolute margins, reflexed. Stamens inserted at the throat; anthers 1.2–1.5 mm. long, ± sagittate; filaments 2.5–4.0 mm. long. Ovary c. 1 mm. long, ovoid; style 3.0–4.5 mm. long, branched at 2.0–3.5 mm., branches 0.8–1.4 mm. long. Fruit c. 4 × 5 mm., subglobose; pyrenes 4, 1-seeded, rugose-sulcate.

Zimbabwe. E: Mutare Distr., Inyamatshira Mt. Range, fl. 22.x.1951, *Chase* 4141 (BM; K; LISC; SRGH). **Malawi**. C: Dedza Mt., fl. & fr. immat. 7.xi.1967, *Salubeni* 868 (K; LISC; SRGH). S: Blantyre Distr., Limbe, Bangwe Hill, fr. 23.xi.1977, *Brummitt, Seyani & Banda* 15167 (K; SRGH). **Mozambique**. MS: Chicamboge *Swynnerton* 1416 (BM; K; SRGH).
    Also recorded from Ethiopia, Zaire, Uganda, Kenya and Tanzania. In the Flora Zambesiaca area apparently confined to the Central and Southern mountainous regions of Malawi and to the mountain range between Zimbabwe (E) and Mozambique (MS). On forest edges, in mixed evergreen forests, rain forests and on rocky gullies; 900–1500 m.

2. **Ehretia amoena** Klotzsch in Peters, Reise Mossamb. Bot.: 248, t. 41 (1861). —Gürke in Engl.,
   Pflanzenw. Ost-Afr. **C:** 335 (1895). —C.H. Wright in F.C. **4**, 2: 5 (1904). —Baker & Wright in
   F.T.A. **4**, 2: 24 (1905) pro parte. —Codd, Trees & Shrubs Kruger Nat. Park: 161, t. 149 (1951).
   —Brenan in Mem. N.Y. Bot. Gard. **9**: 4 (1954). —J.H. Ross, Fl. Natal: 297 (1972). —Drumm. in
   Kirkia **10**: 271 (1975) excl. syn. —Compton, Fl. Swaziland: 480 (1976). —Hall-Martin & Drumm.
   in Kirkia **12**: 177 (1980). Type: Mozambique, Sena, *Peters* s.n. (B†, holotype).
   Ehretia mossambicensis Klotzsch, tom. cit.: 249, t. 42 (1861). —Gürke, loc. cit. (1895). —Schinz &
   Junod in Mém. Herb. Boissier **10**: 60 (1900). —Schinz in Denkschr. Akad. Wiss. Wien, Math.-
   Nat. Kl. **78**: 436 (1905). Type: Mozambique, Rios de Sena, *Peters* s.n. (B†, holotype).
   Ehretia stuhlmannii Gürke, tom. cit.: 336 (1895); in Bot. Jahrb. **28**: 309 (1900). —Baker &
   Wright, tom. cit.: 27 (1905). —Brenan & Greenway, T.T.C.L.: 77 (1949). Syntypes from Tanzania.
   ? *Ficus obovata* T. Sim, For. Fl. Port. E. Afr.: 101, t. 94 fig. C (1909). Type: Mozambique,
   Maganja da Costa, near Mucubela, *Sim* 6018 (K, isotype).

A shrub or sometimes a small tree up to 6(8) m. high, deciduous; branches pubescent to
subglabrous, glabrescent and greyish. Petiole 0.3–1.0(1.7) cm. long, pubescent to
subglabrous; leaf lamina 4–12 × 3–8 cm., circular to oblanceolate or obovate, rough
above, pubescent beneath, rounded to shortly acuminate at apex or, sometimes, truncate
and coarsely crenate-dentate, usually acute or cuneate, rarely obtuse at base, with margins
entire to coarsely crenate-dentate to the upper part, papery to almost leathery, with 5–6
secondary nerves on each side of the midrib, net-veined. Flowers ⚥, 5-merous, a few times
mixed with some 6-merous ones, short-lived, arranged in panicles often corymbiform, up
to 10 × 14 cm., terminal on the shoots or often on very short lateral branches, either
before or with the new leaves; peduncle, rhachis and branches sparsely to densely
pubescent; bracts up to 5 mm. long, subulate, often only a few on the peduncle, sometimes
also on the rhachis and on the branches; pedicels 0.2–1.5(2.5) mm. long, articulated near
apex. Calyx 2.0–3.0 mm. long, lobed to below middle, sometimes to near base, pubescent
outside; lobes ovate to narrowly triangular, acute. Corolla rotate or subrotate, white; tube
(1.5)2.0–3.0 mm. long, widely funnel-shaped, glabrous; lobes longer than the tube,
(2.5)3.0–4.5 × 1.5–2.5 mm., oblong-ovate, obtuse or rounded at apex, ciliate, reflexed.
Stamens about as long as the corolla lobes; anthers 1.5–1.8 mm. long, oblong; filaments
2.5–4.0(4.5) mm. long, slender. Ovary c. 1 mm. long, glabrous; style 3–5 mm. long, divided
$\frac{1}{5}-\frac{2}{5}$, very rarely half-way down; stigmas small, subumbilicate. Fruit c. 5 × 6 × 5 mm.,
subglobose, glabrous; pyrenes 4, with the out surface irregularly foveate.

Zambia. C: Luangwa (Feira), fl. 5.xii.1968, *Fanshawe* 10471 (K). E: Mwangazi valley, fl.
26.xi.1958, *Robson* 730 (BM; K; LISC; SRGH). **Zimbabwe.** N: Mudzi Distr., Mkoto Reserve, airstrip, fl.
& fr. 9.xii.1974, *Müller* 2248 (K; SRGH). E: Mutare Commonage, fl. 16.xi.1948, *Chase* 1271 (BM; K;
SRGH). S: Mwenezi (Nuanetsi) Distr., Litschani's Kraal, Rundi R.,fl. & fr. xi.1956, *Davies* 2202 (K;
LISC; SRGH). **Malawi.** S: Kasupe Distr., Lake Chilwa plain, Chikala Hill, fl. & fr. 28.xi.1973, *Banda*
1198 (SRGH). **Mozambique.** N: Nampula (Moçambique) Prov., between Imala and Muíte, fl.
24.x.1948, *Barbosa* 2558 (LISC; LMA). Z: 67 km. from Alto Molócuè to Gilé, fr. 19.xii.1967, *Torre &
Correia* 16607 (C; FHO; LISC; LMU). T: Cahobra (Cohora) Bassa, 4.5 km. from the dam to Meroeira,
fr. 7.ii.1973, *Torre, Carvalho & Ladeira* 19029 (LD; LISC; LMA; MO; WAG). MS: Sussundenga
(Manica pro parte), Dombe, Matindire, fr. 17.xi.1965, *Pereira & Marques* 672 (LMU). GI: Bilene,
between Chissano and Licilo, fr. 31.iii.1959, *Barbosa & Lemos* 8426 (COI; K; LISC; LMA;
SRGH). M: Matutuíne (Maputo), Salamanga, fl. & fr. 26.xi.1947, *Mendonça* 3527 (C; LISC; LMU; M).
   Also known from Kenya, Tanzania, Swaziland and S. Africa (Transvaal and Natal). In riverine
gallery forests, open forests, woodlands, thickets, cultivated fields and open grounds, often on
sandy-alluvial soils or sometimes on rocky soils; up to 1150 m.
   See note under *E. obtusifolia*.

3. **Ehretia rigida** (Thunb.) Druce in Report Bot. Exch. Club Brit. Is. **1916**: 620 (1917). —Codd, Trees
   & Shrubs Kruger Nat. Park: 162 (1951). —Friedrich-Holzhammer in Merxm., Prodr. Fl. SW. Afr.
   119: 3 (1967). —J.H. Ross, Fl. Natal: 297 (1972). —Palmer & Pitman, Trees of Southern Afr. **3**:
   1943 cum tab. et 3 photogr. (1973). —R.B. Drumm. in Kirkia **10**: 271 (1975). —Compton, Fl.
   Swaziland: 480 (1976). —Palgrave, Trees of Southern Afr.: 803 cum tab. (1977). TAB. **23**. Type
   from S. Africa (Cape).
   Capraria rigida Thunb., Prodr. Pl. Cap.: 103 (1800). Type as above.
   Ehretia hottentotica Burch., Trav. Int. S. Afr. **2**: 147 (1824). —DC., Prodr. **9**: 508 (1845). —Schinz
   & Junod in Mém. Herb. Boiss. **10**: 59 (1900). —C.H. Wright in F.C. **4**, 2: 5 (1904). —Sim, For. Fl.
   Port. E. Afr.: 90 (1909). —Steedman in Proc. Rhod. Sci. Ass. **24**: no. 43 (1925). —Pole-Evans in Fl.
   Pl. S. Afr. **9**: t. 353 (1929). —Miller, B.C.L.: 50 (1948). Type from S. Africa (Cape Prov.).

A small tree up to 6 m. high or more often a many-stemmed shrub 0.8–4.5 m. high,
deciduous; branches glabrous, light-grey to dark-grey. Leaves arranged either alternately
on long juvenile shoots or crowded at the ends of short shoots, these sometimes only

Tab. 23. EHRETIA RIGIDA. 1, flowering twig (×½); 2, flower (× 3), 1–2 from *Barbosa & Lemos* 7759; 3, fruit (× 3); 4, pyrenes (× 3), 3–4 from *Brown* 7574.

vestigial. Petiole up to 0.4 cm. long or absent, very rarely up to 0.8 cm. long. Lamina 1.2–4.0(8.0) × 0.5–2.5(4.0) cm., obovate or sometimes varying from ovate to oblanceolate, sometimes shortly hispid on both surfaces, more often glabrous except by some rigid acroscopic appressed hairs on the margin or quite glabrous, apex rounded or rarely subacute, base narrowly tapering into a short petiole, margins entire, papery to leathery, with 3–4 secondary nerves on each side of the midrib, usually not prominent, sometimes hardly perceptible, often with sparingly pilose acarodomatia at the axils. Flowers ♂, 5-merous, arranged in corymbs 1–3(4) cm. long, usually few-flowered, sessile or shortly pedunculate, puberulous or shortly hispid, terminal on the shoots or on the short-shoots; pedicels 0.2–2.0(4.0) mm. long. Calyx 1.5–2.5 mm. long, variably lobed from above middle to near base, pubescent to subglabrous outside and inside; lobes narrowly triangular to oblong, acute to rounded at apex, ciliate. Corolla mauve to pale blue or white or, a few times purple; tube 3–4(6) mm. long, cylindrical or narrowly funnel-shaped, glabrous; lobes 2.5–4.0 × 1.5–2.0 mm., ovate to oblong, rounded at apex, ciliate, spreading. Anthers 1.0–1.5 mm. long, sagittate-oblong; filaments 2.5–3.5(5.0) mm. long, slender. Ovary glabrous; style 4.0–6.5 mm. long, divided $\frac{1}{5}$–$\frac{2}{5}$-way down; stigmas truncate, sometimes umbilicate. Fruit c. 5 × 7 mm., fleshy, glabrous, orange to red when ripe; pyrenes 4 with the out surface irregularly reticulate-foveate.

Botswana. N: between Lake Ngami and Kgwebe Hills, fr. 17.ii.1966, *Drummond* 8764 (SRGH). SW: 53 km. W. of Kalkfontein, along road to Mamuno, fl. 17.ix.1976, *Bergstrom* B-12 (SRGH). SE: Content Farm, fl. 7.ix.1977, *Hansen* 3177 (SRGH). **Zimbabwe**. W: Between Plumtree and Syringa, fl. & fr. immat. 12.xii.1950, *Orpen* 103/50 (K; SRGH). C: Gweru Distr., 1! km. SE. from Gweru, fr. immat. 29.xii.1966, *Biegel* 1604 (K; SRGH). E: Chipinge Distr., Rupisi, fr. immat. 17.ii.1960, *Farrell* 148 (SRGH). S: Masvingo, *Monro* 2211 (BM). **Mozambique**. GI: Massingir, 23 km. from Lagoa Nova to Chimai, R. Elefantes, fl. & fr. immat. 17.xi.1970, *Correia* 2002 (LMU). M: Matutuíne (Maputo), 5 km. from Bela Vista to Catuane, fl. 23.xi.1948, *Gomes e Sousa* 3887 (COI; K; LMA).

Also known from Angola, Namibia, S. Africa, Swaziland and Lesotho. In savanna woodlands, savannas with shrubs and trees, shrublands and thickets, on granite hills, Kalahari Sands and on alluvial clays, sometimes on termite mounds; up to 1400 m.

See note under *E. obtusifolia*.

4. **Ehretia obtusifolia** Hochst. ex DC., Prodr. **9**: 507 (1845). —Baker & Wright in F.T.A. **4**, 2: 25 (1905). —Fanshawe in For. Res. Bull. **22**: 17 (1973). Type from Ethiopia.
    Ehretia coerulea Gürke in Bot. Jahrb. **28**: 312 (1900). —Baker & Wright, tom. cit.: 24 (1905). —Miller, B.C.L.: 50 (1948). —Brenan & Greenway, T.T.C.L.: 77 (1949). —Wild in Rhod. Agric. Journ. **49**: 288 (1952) "caerulea". —F. White, F.F.N.R.: 364 (1962) *"caerulea"*. —Simpson in Kirkia **10**: 219 (1975). Type from Tanzania.
    Ehretia amoena sensu Baker & Wright, tom. cit.: 24 (1905) pro parte quoad specim. *Lugard 36 et Mrs. Lugard* 48. —Wild, Fl. Vict. Falls: 155 (1953). —Friedrich-Holzhammer in Merxm., Prodr. Fl. SW. Afr. **119**: 3 (1967). —Palgrave, Trees of Southern Afr.: 802, cum tab. (1977) pro parte, non Klotzsch.
    Ehretia mossambicensis sensu N.E.Br. in Kew Bull. **1909**: 122 (1909) pro parte quoad specim. *Lugard 36 et Mrs. Lugard* 48. —Miller, loc. cit. (1948) non Klotzsch.
    Ehretia coerulea var. *glandulosa* Suesseng. in Trans. Rhod. Sci. Assoc. **43**: 42 (1951). Type: Zimbabwe, Marondera, *Dehn* 697 (M, holotype).

Shrub or small much branched tree up to 4.5(6) m. high, deciduous; branches pubescent with short glandular capitate hairs and scattered small bristles to subglabrous, glabrescent and greyish. Petiole 0.4–1.5(2.0) cm. long, glandular-pubescent; leaf lamina 3–8(13) × 1.5–5.0(7.0) cm., obovate or oblanceolate, sometimes widely elliptic, pubescent and with scattered glandular capitate hairs above, pubescent to densely pubescent beneath, sometimes glandular hairs only on the midrib and nerves, rounded to shortly acuminate at apex, cuneate to obtuse at base, with margins entire to coarsely dentate at apex, a few times crenate or crenate-dentate, papery, with 4–6 secondary nerves on each side of the midrib, usually without acarodomatia at the axils. Flowers ♂, 5-merous, arranged in corymbs 3–7 cm. long, usually up to 30-flowered, rarely more, subsessile to well-pedunculate, terminal on the shoots or on the short shoots and often afterwards seemingly extra-axillary; peduncle and branches glandular-pubescent and sometimes bristly; pedicels 0.5–5.0 mm. long. Calyx 1.5–2.5 mm. long, lobed to near base, glandular-pubescent; lobes narrowly triangular to ovate, acute to obtuse at apex, ciliate. Corolla blue to mauve, rarely white or purple; tube 3.5–5.5 mm. long, cylindrical to narrowly funnel-shaped; lobes 3–4 × 1.5–2.0 mm., ovate to oblong, rounded at apex, ± spreading. Anthers 1.0–1.6(1.8) mm. long, sagittate-oblong to linear; filaments 3.0–4.5 mm. long. Ovary ovoid, glabrous or glandular-puberulous; style 5–7 mm. long, divided from $\frac{1}{10}$–$\frac{1}{3}$ of its length. Fruit

c. 5 × 7 × 5 mm., fleshy, often glandular-pubescent, sometimes glabrous, orange when ripe; pyrenes 4, irregularly reticulate-foveate on the outer surfaces.

Botswana. N: Linyanti, Serondella, fl. xii.1950, *Miller* B/1136 (K; PRE). SE: 25 km. NNE. of Palapye on road towards Serule, fl. 1.x.1978,*Hansen* 3476 (SRGH). **Zambia**. W: Kitwe, fr. 26.i.1958, *Fanshawe* 4201 (FHO; K). C: 9.5 km. E. of Lusaka, fl. 20.xii.1957, *King* 395 (K). S: Choma Distr., Mapanza, fl. 23.xi.1957, *Robinson* 2505 (K; PRE; SRGH). **Zimbabwe**. N: Rushinga Distr., near Nyaderi, Pfungwe Reserve, fl. 18.x.1955, *Lovemore* 449 (K; LISC; SRGH). W: Matobo, hillside above houses R.M.E., fl. & fr. 7.xi.1947, *West* 2434 (K; SRGH). C: 6.5 km. E. from Gweru, fl. 25.xi.1966, *Biegel* 1474 (K; SRGH). E: Mutare Distr., Hawkhead Farm, fl. & fr. immat. 12.xi.1950, *Chase* 3071 (BM; K; LISC; SRGH). S: Masvingo Distr., near Great Zimbabwe, fr. immat. 6.xii.1960, *Leach & Chase* 10563 (BM; LISC; SRGH). **Malawi**. N: Chitipa Distr., 32 km. SE. of Chisenga, fl. 3.i.1977, *Pawek* 12216 (SRGH). S: Mangochi–Monkey Bay Rd., fl. 25.xi.1954, *Jackson* 1403 (FHO; K). **Mozambique**. T: 5 km. from Tete to Changara, fl. 21.xii.1965, *Torre & Correia* 13775 (COI; K; LISC; LMU). M: Between Boane and Goba, fr. 16.ii.1949, *Myre & Balsinhas* 360 (LMA; SRGH).

Known from Ethiopia and through east Africa southwards to South Africa (Transvaal), Angola and Namibia. Also recorded from Afghanistan, Pakistan and India. In woodlands, thickets, granite hills, very often on termite mounds; 150–1500 m.

E. amoena, E. rigida and *E. obtusifolia* are widespread and very closely related. For the most part these are readily distinguished but in areas where the species are sympatric specimens with intermediate features occur. Thus specimens from Botswana (Lake Ngami and Kgwebe Hills (N)), Zimbabwe (Matopos and Bulawayo (W), Gweru (C), Mutare (E), Masvingo (S)) and Mozambique (Namaacha and Goba (M)) reveal features either of *E. rigida* or *E. obtusifolia*, both co-existing in these regions. Likewise, specimens gathered in Zambia (Luangwa Valley), Zimbabwe (Mutare, Chimanimani and Chipinge (E)), Malawi (Blantyre, (S)) and in Mozambique (Tete (T) and Chemba (MS)) show features of either *E. amoena* (incl. *E. stuhlmannii*) or *E. obtusifolia* or intermediate features. Specimens from Southern Mozambique (Limpopo river (MS) and southwards to Natal border), show features intermediate between *E. amoena* and *E. rigida*. These specimens exhibiting intermediate features and morphological diversity are, most probably, hybrids.

4. COLDENIA L.

Coldenia L., Sp. Pl.: 125 (1753); Gen. Pl. ed. 5: 61 (1754).

Herbs with usually procumbent branched stems. Leaves alternate, usually numerous, crenate or lobate, subsessile or petiolate. Flowers 4-merous, solitary, extra-axillary, the upper ones in leafy spicate branches. Calyx deeply lobed. Corolla actinomorphic, small, naked. Stamens inserted below the middle of the corolla tube, included. Nectary absent. Ovary 4-celled with 1 ovule pendulous in each cell. Styles 2, slightly united below, terminal; stigmas punctate. Fruit dry or slightly succulent, ± ovoid-pyramidal, 4-lobed, usually separating into 4, 1-seeded nutlets. Seeds ovoid.

A monotypic genus occurring in the tropical regions of Africa, Asia and Australia.

Coldenia procumbens L., Sp. Pl.: 125 (1753). —Gürke in Engl., Pflanzenw. Ost-Afr. C: 336 (1895). —Hiern, Cat. Afr. Pl. Welw. **1**: 717 (1898). —Baker & Wright in F.T.A. **4**, 2: 28 (1905). —Taton in Fl. Congo, Rwanda et Burundi, Boraginaceae: 26, t. 3 (1971). TAB. **24**. Type from India.

A procumbent annual herb; stems up to 40 cm. long, radiating, angular or ± laterally compressed, much branched, densely hairy to villous. Leaves with petiole 1–5(12) mm. long, on the first ones up to 40 mm. long; lamina 4–23 × 3–15 mm. (on the first leaves 25–43 × 15–32 mm.) usually oblong, sometimes circular, ovate, elliptic or obovate, plicate on the youngest leaves, with appressed hairs diverging from the nerves and converging to the teeth apex above, densely glandular-hairy below, apex rounded, base asymmetrical, cuneate on one side, obtuse to truncate on the other, margins deeply crenate to lobed. Flowers subsessile. Calyx 2.5–3.0 mm. long, a little accrescent in fruit, villous outside and inside; lobes ovate-lanceolate, subequal. Corolla 1.5–2.2 mm. long, narrowly conical, calyptriform, early deciduous, white, glabrous. Stamens inserted at c. 0.5 mm. from the base of the corolla tube; anthers c. 0.3 mm. long, suborbicular to broadly elliptic; filaments as long as the anthers, glabrous. Ovary c. 0.4 mm. long, ovate-pyramidal, glandular-pubescent; styles slightly united at base, early concealed by the apical protuberances of the fruit. Fruit 4–5 mm. wide, depressed-ovoid, 4-lobed, beaked and with irregular protuberances, glandular-hairy, brownish, dividing at first into 2 pairs of nutlets, later into 4 nutlets with the ventral surface angular.

80

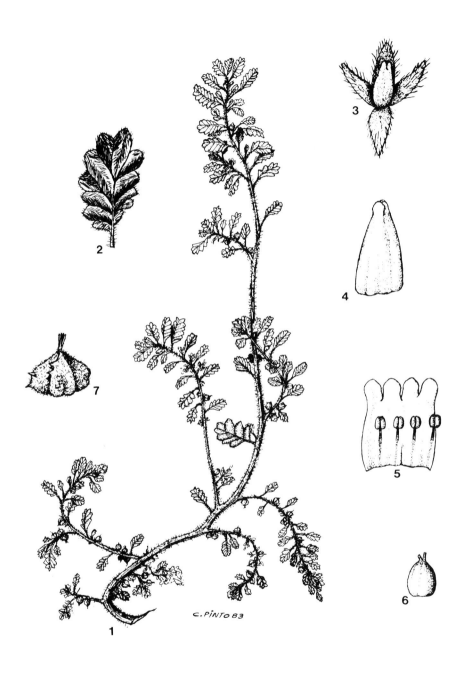

Tab. 24. COLDENIA PROCUMBENS. 1, part of stem (× ½); 2, leaf upper surface (× 2); 3, flower (× 5); 4, corolla (× 10); 5, corolla opened out to show stamens (× 10); 6, gynoecium (× 10); 7, fruit (× 3), all from *Guy* 857.

Zambia. C: Luangwa R., Mfuwe lagoon, fl. & fr. immat. 27.vi.1964, *Vesey-FitzGerald* 4308 (K). E: Nsefu Game Camp, fl. & fr. 15.x.1958, *Robson & Angus* 113 (BM; K; LISC; SRGH). **Zimbabwe**. N: Kariba Distr., Sengwa, fl. & fr. 18.xi.1964, *Jarman* 88 (K; SRGH). **Mozambique**. MS: Gorongosa Nat. Park, Urema flood-plains, fl. & fr. vi.1970, *Tinley* 1947 (K; LISC; SRGH).

Widespread in tropical Africa; also in Madagascar, tropical Asia and tropical Australia. In plains, banks and edges of rivers and lagoons on sandy or muddy soils seasonally flooded; 90–750 m.

5. ARGUSIA Boehmer

Argusia Boehmer in Ludwig, Defin. Gen. Pl., ed. Boehmer: 507 (1760).
Messerschmidia L. ex Hebenstr. in Nov. Comment. Acad. Sci. Imp. Petrop. **8**: 315 (1763).

Small trees, shrubs or perennial herbs. Leaves entire, sessile or shortly petiolate. Inflorescence of scorpioid ebracteate cymes arranged in terminal corymbs or panicles. Flowers 5-merous, sessile or pedicellate. Calyx 5-lobed almost to the base. Corolla actinomorphic, infundibuliform or hypocrateriform; limb 5-lobed to a half or more. Stamens included, sometimes reaching as far as the throat or a little more, free. Nectary absent or minute. Ovary 4-celled, ovules solitary, pendulous. Style terminal; stigmatic ring topped by an entire to 2-lobed stigmatic appendix. Fruit dry, dividing at maturity into two 2-seeded mericarps; mesocarp corky; the two fertile cells of each mericarp separated by a deep groove or a sterile cavity.

A small genus comprising c. 5 species and occurring in the SE. of the United States of America, Central America and West Indies, Canary Is., SE. Europe, Temperate Asia and in the isles of the Indian and Pacific Oceans.

Argusia argentea (L.f.) Heine in Fl. N. Caléd. et Dépend. **7**: 109, t. 24 (1976). TAB. **25**. Type from Sri Lanka.
Tournefortia argentea L.f., Suppl. Pl.: 133 (1781). —DC., Prodr. **9**: 514 (1845). —Klotzsch in Peters, Reise Mossamb. Bot.: 250 (1861). —Baker, Fl. Maurit.: 201 (1877). —Gürke in Engl., Pflanzenw. Ost-Afr. **C**: 336 (1895). —Baker & Wright in F.T.A. **4**, 2: 29 (1905). Type as above.
Messerschmidia argentea (L.f.) I.M. Johnston in Journ. Arnold Arbor. **16**: 164 (1935). Type as above.

Erect shrub or small tree up to 4 m. high; branches stout with prominent leaf scars, densely hairy or tomentose. Leaves spirally arranged and crowded at ends of the branches, 7–22 × 4–11 cm., ovate to obovate or oblanceolate, sometimes rhombic, greyish or brownish, velvety on both surfaces, rounded or obtuse as apex, gradually or suddenly narrowed to the base, sessile or with a petiole up to 2 cm. long. Cymes arranged in ample long-pedunculate terminal panicle or corymb with the peduncle, rhachis and branches ± angular or sometimes subwinged. Flowers numerous, sessile. Calyx 2.0–3.0 mm. long, velvety outside, glabrous inside, persisting after the fruits have fallen; lobes broadly ovate to elliptic, rounded at apex. Corolla tube as long as the calyx or sometimes a little shorter, pubescent to glabrous outside, glabrous inside, greenish; corolla lobes imbricate in bud, 1.2–1.4 × 1.0–1.4 mm., circular to oblong, rounded at apex, hairy towards the base outside, glabrous inside, white. Stamens inserted about the middle of the corolla tube; anthers c. 1.5 mm. long, ovate-lanceolate, cordate at the base, reaching as far as the throat or a little more, subsessile. Ovary c. 1 mm. long, broadly conical, ridged, glabrous. Style subnull; stigmatic ring c. 0.7 mm. in diam., thick, ridged, topped by the stigmatic appendix up to 1 mm. long, cleft to the base. Fruit c. 5 × 7 mm., subglobose, remotely ribbed to the apex, grey-brownish, glabrous; mesocarp vesicular, corky, much developed in the lower $\frac{2}{3}$ of the fruit; mericarps with horny endocarp, bilocular, 2-seeded, with a deep and narrow fissure between the cells or sometimes with an empty vestigal cavity between them.

Mozambique. N: Caldeira I., fl. & fr. 27.x.1965, *Mogg* 32363 (LISC; SRGH). Z: Fogo I., fl. & fr. 14.xi.1952, *Gomes e Sousa* 4027 (K; LISC). GI: Inhambane, Pomene, fl. xii.1971, *Tinley* 2264 (K; LISC; SRGH).

In Africa also known from Kenya and Tanzania. Also recorded from the isles and islets of the Indian and Pacific Oceans, mainly those of coraline origin. On rocky coral-beaches.

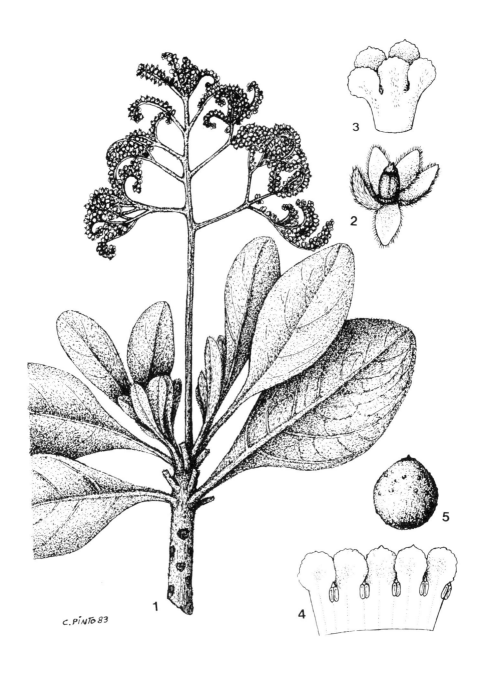

Tab. 25. ARGUSIA ARGENTEA. 1, flowering twig (× ½); 2, calyx and gynoecium (× 6); 3, corolla (× 6); 4, corolla opened out to show stamens (× 6); 5, fruit (× 3), all from *Mogg* 32363.

6. HELIOTROPIUM L.

Heliotropium L., Sp. Pl.: 130 (1753); Gen. Pl. ed. 5: 63 (1754).

Annual or perennial herbs, sometimes woody at the base, rarely shrubs. Leaves alternate or rarely subopposite, entire or denticulate. Cymes spike-like, usually scorpioid, terminal or pseudoaxillary, bracteate or ebracteate, simple or branched or sometimes flowers solitary supra-axillary. Flowers small, 5-merous, hermaphrodite. Calyx usually lobed to near base. Corolla hypocrateriform or infundibuliform, white, yellow or blue; lobes usually spreading, imbricate or induplicate in bud. Stamens included; filaments very short. Ovary completely or incompletely 4-celled; ovules pendulous; style terminal, usually short, included, with a stigmatic ring below the apex and usually with a cylindrical or conical terminal sterile appendix. Fruit dry, usually splitting at maturity into 4, 1-seeded nutlets, free or cohering in pairs or remaining entire and 1-seeded by abortion; seeds usually with a thin endosperm.

A genus of about 280 species, occurring in temperate and warm regions, usually in open habitats.

1. Plants fleshy, glabrous, glaucous - - - - - - - - - 13. *curassavicum*
– Plants not as above - - - - - - - - - - - - 2
2. Corolla lobes acuminate or caudate - - - - - - - - - - 3
– Corolla lobes neither acuminate nor caudate - - - - - - - - 5
3. Anthers apiculate; plants with 2-branched trichomes - - - - - - 7. *lineare*
– Anthers bifid or emarginate at apex; plants without 2-branched trichomes - - 4
4. Calyx lobes thick, callose; fruit tuberculate-crested, saddle-shaped at apex; nutlets conate
    in pairs - - - - - - - - - - - - - 4. *ciliatum*
– Calyx lobes thin, 3-ribbed; fruit tuberculate, subglobose; nutlets free 5. *zeylanicum*
5. Flowers solitary, supra-axillary; fruits beaked - - - - 1. *baclei* var. *rostratum*
– Flowers in terminal or pseudo-terminal cymes; fruits not beaked - - - 6
6. Cymes bracteate - - - - - - - - - - - - 2. *strigosum*
– Cymes not bracteate - - - - - - - - - - - - 7
7. Leaves ovate, broadly ovate to broadly obovate - - - - - - - 8
– Leaves lanceolate, oblong, elliptic or obovate - - - - - - - 10
8. Leaves large, more than 25 mm. wide; fruits mitre-shaped, not enclosed in the calyx 9
– Leaves small, up to 20 mm. wide; fruits not mitre-shaped, enclosed in the
    calyx - - - - - - - - - - - - - - 12. *supinum*
9. Fruit-lobes remarkably divergent; mature nutlets with an empty cavity 9. *indicum*
– Fruit-lobes slightly divergent; mature nutlets without an empty cavity 8. *elongatum*
10. Leaves obovate or, rarely, elliptic, apiculate; style very short; fruit clothed with
    short hairs - - - - - - - - - - - - - 3. *ovalifolium*
– Leaves never obovate, not apiculate; style well developed; fruit glabrous - - - 11
11. Corolla tube with retrorse hairs; fruit subglobose, pustulate, nutlets free 6. *giessii*
– Corolla tube with antrorse hairs; fruit ± obpyriform, not pustulate; nutlets cohering
    in pairs - - - - - - - - - - - - - - 12
12. Calyx to 2.5 mm. long; corolla tube to 2.5 mm. long - - - - 10. *harareense*
– Calyx more than 2.5 mm. long; corolla tube more than 3.5 mm. long 11. *steudneri*

1. **Heliotropium baclei** DC., Prodr. **9**: 546 (1845). —Baker & Wright in F.T.A. **4**, 2: 34 (1905). —I.M. Johnston in Contr. Gray Herb., part 92: 91 (1930). Type from Gambia.

Var. **rostratum** I.M. Johnston, loc. cit. (1930). —Taton in Fl. Congo, Rwanda et Burundi, Boraginaceae: 35 (1971). TAB. **26** fig. 1. Syntypes from Zaire (Katanga).
    Heliotropium baclei sensu Wild, Fl. Vict. Falls: 155 (1953).
    Heliotropium katangense Gürke ex De Wild. in Ann. Mus. Congo, Bot. Sér. 4, **1**: 223 (1903). —Baker & Wright, tom. cit.: 43 (1905). Syntypes as for *H. baclei* var. *rostratum*.
    Heliotropium marifolium sensu Baker & Wright, tom. cit.: 40 (1905) pro parte quoad specim. Scott. —Friedrich-Holzhammer in Merxm., Prodr. Fl. SW. Afr. **119**: 8 (1967) non Retz.

Decumbent diffuse perennial herb. Stems usually slender but sometimes relatively stout, ± densely pubescent to strigose. Leaves alternate; petiole 1–4(6) mm. long, pubescent to strigose; lamina 4–18(22) × 2–8 mm., elliptic to oblanceolate, rarely obovate, strigose or appressed-pubescent above and below, with the swollen bases of the hairs surrounded by cystolithic cells, subacute to obtuse at apex, acute at base, margins entire, secondary nerves inconspicuous. Flowers solitary, supra-axillary. Pedicels 0.5–2.0 mm. long, hairy. Calyx sparsely strigose or pubescent outside, strigose to glabrous inside; lobes 2.0–5.0 × 0.5–1.5 mm., subequal or unequal (rarely 1 lobe up to 10 × 3 mm.), lanceolate, acute, ciliated to the base. Corolla infundibuliform, bright yellow, with some bristly hairs

84

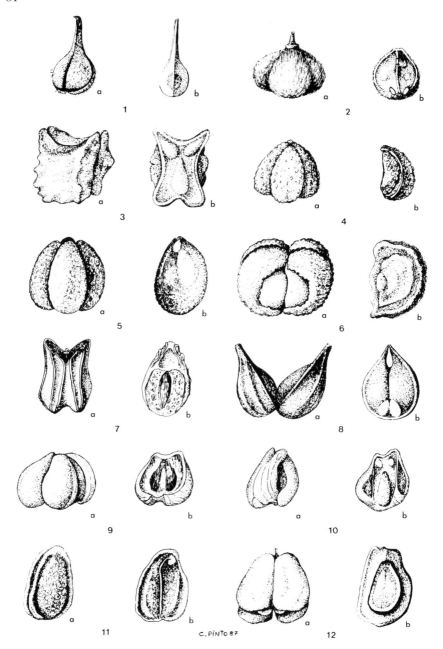

Tab. 26. HELIOTROPIUM, a = fruit, b = nutlets. 1. —H. BACLEI var. ROSTRATUM, (× 4) from *van Rensburg* 2753. 2. — H. OVALIFOLIUM, (× 10) from *Stewart* 12. 3. —H. CILIATUM, (× 4) from *Pedro & Pedrogão* 1501. 4. —H. ZEYLANICUM, (× 8) from *Wild* 3447. 5. —H. GIESSII, (× 8) from *Mavi* 1524. 6. —H. LINEARE, (× 8) from *Van Son* in H.T.M. 28780. 7. —H. ELONGATUM, (× 5) from *Blackmore & Patel* 287. 8. —H. INDICUM, (× 4) from *Barbosa* 8300. 9. —H. HARAREENSE, (× 5) from *Greatrex* in GHS 26562. 10. —H. STEUDNERI, (× 5) from *de Hoog* 83. 11. —H. SUPINUM, (× 5) from *Davies* 1383. 12. —H. CURASSAVICUM, (× 8) from *Mendonça* 120.

outside, glabrous inside save by 5 fringed minute scales at the level of the anthers; tube 3.5–5.0 mm. long; lobes 0.7–1.5 × 0.7–1.3 mm., broadly triangular with apex obtuse and often bent inwards, usually with 5 minute lobes or teeth alternating with them. Stamens inserted at c. 1 mm. from the base on the corolla tube; anthers 0.7 mm. long, ovate, appendiculate, cohering by the puberulous appendages around the sterile stigmatic appendix, subsessile. Ovary ovoid, glabrous or subglabrous. Style 0.2–0.3 mm. long, glabrous; stigmatic ring not well-marked, topped by the sterile appendix 0.4–0.6 mm. long, conical, 2–4-fid at apex. Fruits 4–6 × 2–3 mm. (including the 2–3 mm. long terminal beak), ovoid-rostrate, with some scattered rigid hairs, usually some of them crozier-shaped; beak glabrous or subglabrous; nutlets 4, free.

Caprivi Strip. Katima Mulilo area, fl. & fr. 24.xii.1958, *Killick & Leistner* 3063 (K; SRGH). **Botswana**. N: Chobe Nat. Park, fl. & fr. 28.x.1969, *Mahundu* CNP/1/61 (SRGH). **Zambia**. B: Mongu Distr., Sandaula Pontoon, fl. & fr. 9.xi.1959, *Drummond & Cookson* 6264 (K; LISC; SRGH). N: Kawambwa, fl. & fr. 10.xi.1957, *Fanshawe* 3887 (K; LISC). W: Kasempa Distr., Lufupa R. north of Ntembwa, fl. & fr. 24.ix.1964, *van Rensburg* 2962 (K; SRGH). C: Lusaka Distr., Lake Kafue, below Kafue Bridge, fl. & fr. 26.xii.1972, *Kornaś* 2856 (K). S: Mazabuka Distr., Kafue Gorge, fl. & fr. 24.xi.1959, *Drummond & Cookson* 6762 (K; LISC; SRGH). **Zimbabwe**. N: Hurungwe Distr., near Chirundu Bridge, fl. & fr. 8.xii.1961, *Whellan* 1884 (K; SRGH). W: Hwange Distr., Victoria Falls, fl. & fr. 21.xii.1978, *Mshasha* 145 (SRGH). C: Harare Distr., Spelonken, Mazowe (Mazoe) Dam, fl. x.1980, *Burrows* 1463 (K; SRGH). **Malawi**. S: Elephant Marsh, fl. & fr. xii.1887, *Scott* (K).
Also recorded from Mali, Central African Republic, Zaire, Tanzania, Angola and Namibia. On riverbanks, streamsides and in seasonally flooded grounds, on muddy or sandy soils; 800–1200 m.

2. **Heliotropium strigosum** Willd., Sp. Pl. **1**: 743 (1798). —DC., Prodr. **9**: 546 (1845). —Hiern, Cat. Afr. Pl. Welw. **1**: 719 (1898) pro parte quoad specim. Andong. —Baker & Wright in F.T.A. **4**, 2: 41 (1905) pro parte excl. saltem specim. *Welwitsch* 5299 & *Baum* 8. —N.E. Br. in Kew Bull. **1909**: 122 (1909). —Friedrich-Holzhammer in Merxm., Prodr. Fl. SW. Afr. **119**: 9 (1967). —Binns, H.C.L.M.: 24 (1968). —Hilliard & Burtt in Notes Roy. Bot. Gard. Edin. **30**: 117 (1970). —Taton in Fl. Congo, Rwanda et Burundi, Boraginaceae: 34 (1971). —Hall-Martin & Drumm. in Kirkia **12**: 177 (1980). TAB. **27**. Type from W. Africa.
Heliotropium longifolium Klotzsch in Peters, Reise Mossamb. Bot.: 251 (1861). Type: Mozambique, Rios de Sena, *Peters* s.n. (B†, holotype).
Heliotropium pygmaeum Klotzsch, op. cit.: 252 (1861). Type: Mozambique, near Tete, *Peters* s.n. (B†, holotype).
Heliotropium senense Klotzsch, op. cit.: 253 (1861). Type: Mozambique, Rios de Sena, *Peters* s.n. (B†, holotype).
Heliotropium constrictum Kaplan in Ann. Transv. Mus. **12**: 186, t. 3 (1927). Type: Zambezi R., *Wilde* in H.T.M. 9058 (PRE, holotype).

Erect annual or sometimes perennial herb, 15–35 cm. high; stem and branches ± angular, subterete to the base, strigose. Leaves alternate; petiole 0.5–3.0(9) mm. long, strigose; lamina 10–35(50) × 1–6(9) mm., linear-lanceolate, strigose on both surfaces, apex and base acute, margins entire usually slightly revolute. Cymes simple, often numerous, lax, 4–10(14) cm. long when wholly expanded, terminal; rhachis strigose; bracts like the leaves, the lower ones 5–11 × 1–3 mm., gradually smaller to the cyme end. Flowers with pedicels 0.3–0.5 mm. long, a little accrescent, angular, decurrent on the rhachis. Calyx 1.5–2.5 mm. long, lobes ovate-lanceolate, ciliate, strigose outside, glabrous or subglabrous inside, persisting after the nutlets are shed. Corolla 2.5–3.5(4) mm. long, infundibuliform, strigose on the upper $\frac{2}{3}$ outside, glabrous inside except for 5 tufts of small flattened hairs at the level of anther tips, white or with orange-yellow tube; lobes 0.5–1.0 × 0.7–1.5 mm., very widely triangular-ovate, obtuse or rounded, erect. Stamens inserted near the base on the corolla tube, anthers 0.7–0.8 mm. long, ovate-lanceolate, cohering by the minutely pubescent tips; filaments 0.2–0.3 mm. long. Ovary glabrous; style 0.2–0.4(0.5) mm. long, glabrous; stigmatic ring c. 0.4 mm. in diam.; sterile appendix 0.2–0.5 mm. long, narrowly conical. Fruit up to 1.4 × 2.2 mm., depressed, shortly strigose on the upper $\frac{1}{2}$ or $\frac{2}{3}$, late splitting into 4 free nutlets; nutlets ovoid, dark brown to black when mature, rounded on the back, with a deep central-peripheric hollow on each of the inner surfaces.

Botswana. N: Ngamiland, Okavango, near Tsau, fl. & fr. 8.iii.1961, *Richards* 14782 (K; SRGH). SW: Ghanzi Distr., fl. & fr. 6.i.1968, *Buerger, Boshoff & Mason* in MMC 218 (K). **Zambia**. C: Katondwe, fl. & fr. 4.ii.1964, *Fanshawe* 8288 (K; SRGH). **Zimbabwe**. N: Guruve (Sipolilo) Distr., Kanyemba Camp, fl. & fr. 1.ii.1966, *Müller* 319 (K; SRGH). W: Bulilima Mangwe, Lupandeni, fl. & fr. 10.iv.1942, *Feiertag* in GHS 45390 (K; SRGH). E: E. Savi, between Musvasivi (Musaswe) R. and Cikariati R., fl. & fr. 22.i.1957, *Phipps* 113 (K; SRGH). S: Limpopo Ranches Rd., between turn off from Bulawayo Rd. and Umzingwane R., fl. & fr. 25.iii.1959, *Drummond* 6031 (K; LISC; SRGH).

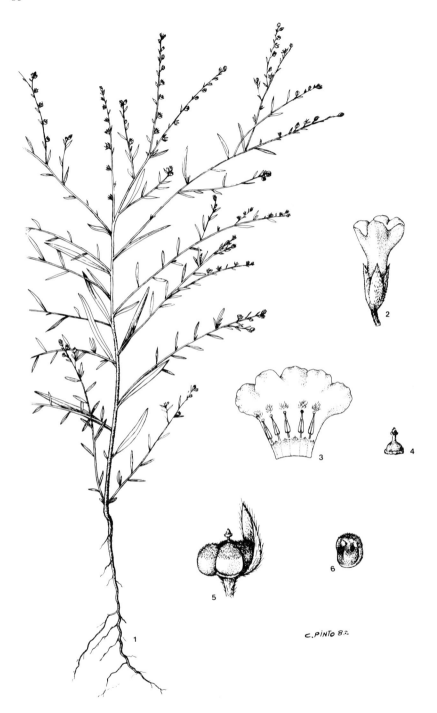

Tab. 27. HELIOTROPIUM STRIGOSUM. 1, habit (× ½); 2, flower (× 8); 3, corolla opened out to show stamens (× 8); 4, gynoecium (× 8), 1–4 from *Torre & Paiva* 9951; 5, fruit (× 8); 6, nutlet (× 8), 5–6 from *Müller* 319.

Malawi. C: Salima, between L. Malawi and Grand Beach Hotels, fl. & fr. 16.ii.1959, *Robson & Steele* 1623 (BM; K; LISC). S: Chikwawa Distr., Lengwe Game Reserve, fl. & fr. 31.i.1970, *Hall-Martin* 522 (K; SRGH). **Mozambique.** N: Nampula, Erati, Namapa, C.I.C.A. Res. St., fl. & fr. 28.iii.1961, *Balsinhas & Marrime* 321 (BM; COI; K; LISC; LMA; LMU; SRGH). Z: Pebane, fl. & fr. 9.iii.1966, *Torre & Correia* 15131 (C; LISC; LMU). T: 3 km. from Tete to Changara, fl. & fr. 13.ii.1968, *Torre & Correia* 17547 (LD; LISC; LMA). MS: Manica, near Bandula, fl. & fr. 12.iii.1948, *Garcia* 576 (LISC; LMU; WAG). GI: Govuro, 3 km. from Banamana to Machaíla, fl. & fr. 19.iii.1974, *Correia & Marques* 4074 (LMU). M: 10 km. from Namaacha to Matianine, fl. & fr. 10.iii.1975, *Marques* 2570 (LMU).

Throughout tropical Africa and in Egypt, Arabia, tropical Asia and Australia. In open woodlands and grasslands on shallow or sandy soils, roadsides, denuded grounds and coastal dunes; up to 1200 m.

3. **Heliotropium ovalifolium** Forssk., Fl. Aegypt.-Arab.: 38 (1775). —Gürke in Engl., Pflanzenw. Ost-Afr. **C**: 337 (1895). —Hiern, Cat. Afr. Pl. Welw. **1**: 718 (1898). —C.H. Wright in F.C. **4**, 2: 8 (1904). —Baker & Wright in F.T.A. **4**, 2: 34 (1905). —N.E. Br. in Kew Bull. **1909**: 122 (1909). —R. Good in Journ. of Bot. **67**, Suppl. **2**: 107 (1929). —Bremek. & Oberm. in Ann. Transv. Mus. **16**: 431 (1935). —Weimarck in Bot. Notis. **1940**: 63 (1940). —Wild in Rhod. Agric. Journ. **50**: 418 (1953). —Brenan in Mem. N.Y. Bot. Gard. **9**: 6 (1954). —Wild, op. cit. **52**: 535 (1955); in Common Rhod. Weeds: fig. 81 (1955). —Rattray & Wild in Rhod. Agric. Journ. **52**: 501 (1955). —Friedrich-Holzhammer in Merxm., Prodr. Fl. SW. Afr. **119**: 9 (1967). —Taton in Fl. Congo, Rwanda et Burundi, Boraginaceae: 32 (1971). —J.H. Ross, Fl. Natal: 297 (1972). —W.B.G. Jacobsen in Kirkia **9**: 171 (1973). —Hall-Martin & Drumm. in Kirkia **12**: 177 (1980). TAB. **26** fig. 2. Type from Arabia.

Heliotropium coromandelianum Retz., Obs. Bot. **2**: 9 (1781). —DC., Prodr. **9**: 541 (1845). Type from India.

Heliotropium coromandelianum var. *ovalifolium* (Forssk.) Lehm., Pl. Asper.: 46 (1818). — Klotzsch in Peters, Reise Mossamb. Bot.: 253 (1861). Type as for *H. ovalifolium*.

Heliotropium apiculatum E. Meyer ex Drège in Flora **26**, 2, Beigabe (Zwei Pfl. Docum.): 191, 93 (1843) nom. tantum.

Heliotropium coromandelianum var. *obovatum* DC., loc. cit. (1845). Type as for *H. coromandelianum*.

Heliotropium phyllosepalum Baker in Kew Bull. **1894**: 30 (1894). —Baker & Wright in F.T.A. **4**, 2: 33 (1905). Type: Mozambique, Morrumbala, banks of Shire R., *Scott* s.n. (K, holotype).

Erect or decumbent annual or perennial herb up to 60(90) cm. high with a thin rootstock; stems pubescent, woody or subwoody to the base, branches often laterally compressed and silvery-sericeous to the extremities. Leaves numerous, alternate; petiole 2–20 mm. long, pubescent to villous; lamina 10–60 × 4–30 mm., obovate or elliptic, pubescent, strigose-villous or tomentose on both surfaces, with hairs of two kinds, long and hardy, appressed and with swollen bases, some weak, short and spreading, at least on the basal leaves, the others, rounded to retuse, rarely subacute but always apiculate or mucronate at apex, acute or subacute at base, margins slightly revolute. Cymes dense, up to 17 cm. long when completely developed, often 2, sometimes 1 or 3, rarely 4 together; common peduncle 2–4(6) cm. long. Flowers subsessile but pedicels increasing up to 2 mm. in fruit and decurrent on the rhachis. Calyx lobes unequal, the largest one up to 3.0 × 1.0 mm. (6.0 × 1.5 mm. in fruit), ovate, the smallest one 1.0–2.0 × 0.2–0.4 mm., lanceolate, strigose outside and inside. Corolla pubescent outside, glabrous inside save by a ring of flattened hairs (sometimes lacking) at the level of the anther tips, usually with yellow tube and white limb; tube 1.5–3.0 mm. long, slightly widened at the middle; lobes 0.5–1.5(2.0) × 0.4–1.5 mm., unequal or subequal, broadly ovate to triangular-ovate, suberect, apex acute or apiculate. Stamens inserted on the lower third of the corolla tube; anthers 0.7–1.2 mm. long, narrowly lanceolate, minutely hairy at apex, subsessile. Style very short; stigmatic ring well-marked; sterile appendix 0.3–0.6 mm. long, conical with apex truncate or minutely 2-dentate, glabrous or shortly hairy. Fruits c. 1.5 × 2.0 mm., subglobular depressed, densely clothed with whitish appressed short hairs; nutlets 4, free, with the inner surfaces without cavities.

Caprivi Strip. About 7 km. from Katima Mulilo to Lisikili, fl. & fr. 24.xii.1958, *Killick & Leistner* 3088 (K; SRGH). **Botswana.** N: Maun, Thamalakane R., fl. & fr. 22.i.1972, *Biegel & Russell* 3726 (K; LISC). SW: Ghanzi Pan, 14 km. E. of Ghanzi, fl. & fr. 21.x.1969, *Brown* 6987 (K; SRGH). SE: Thalamabele-Mosu area, near Sowa Pan, fl. & fr. 14.i.1974, *Ngoni* 327 (SRGH). **Zambia.** B: 16 km. N. of Senanga, fl. & fr. 31.vii.1952, *Codd* 7284 (BM; K; PRE; SRGH). N: Mweru-Wantipa, Kasongole, Katanga border, fl. & fr. 3.viii.1962, *Tyrer* 217 (BM; SRGH). W: Kitwe, fl. & fr. 28.ix.1963, *Fanshawe* 8000 (K). C: Lusaka, fl. & fr. 14.viii.1963, *Robinson* 5597 (K; SRGH). E: Nsefu Game Camp, Luangwa R., fl. & fr. 15.x.1958, *Robson & Angus* 137 (BM; K; LISC; SRGH). S: Mumbwa Distr., Kafue Hoek pontoon, fl. & fr. 21.xi.1959, *Drummond & Cookson* 6749 (K; LISC; SRGH). **Zimbabwe.** N: Hurungwe Distr., Mensa Pan, 17 km. E.SE. of Chirundu Bridge, fl. & fr. 30.i.1958, *Drummond* 5385 (K; LISC). W: Hwange Distr., Lukosi R., E. of Mbala Lodge, fl. & fr. 21.x.1968, *Rushworth* 1212 (K; LISC;

SRGH). C: Chegutu Distr., Umfuli R., fl. & fr. 7.xii.1960, *Rutherford-Smith* 414 (SRGH). E: Vumba Mts., near Mutare (Umtali), fl. & fr. xii.1937, *Obermeyer* 2039 (BM; PRE). S: Ndanga Distr., Rundi R., Chipinda Pools area, fl. & fr. 9.xii.1959, *Goodier* 718 (K; LISC; SRGH). **Malawi**. N: Karonga Distr., Ngala, 30 km. N. of Chilumba, fl. & fr. 1.i.1973, *Pawek* 6263 (K; SRGH). C: near Salima, fr. 16.ii.1959, *Robson & Steele* 1616 (BM; K; LISC; SRGH). S: Mulange, Phalombe Rice Gardens, fl. & fr. 16.xi.1955, *Jackson* 1762 (K; SRGH). **Mozambique**. N: Nampula, Mecuburi, Muíte, R. Lúrio, fl. & fr. 21.xi.1936, *Torre* 1075 (COI; LISC). Z: Mopeia, fl. & fr. 12.ix.1944, *Mendonça* 2038 (C; LISC; LMU; MO; WAG). T: 43 km. from Tete to Chicoa, fl. & fr. 27.xii.1965, *Torre & Correia* 13859 (LISC; LMA). MS: Gorongosa, Chitengo, R. Púngoè, fl. & fr. 23.x.1965, *Balsinhas* 1007 (COI; LMA). GI: Limpopo (Baixo Limpopo), Aldeia da Barragem, fl. & fr. 20.xi.1957, *Barbosa & Lemos* 8228 (COI; K; LISC; LMA). M: Sábiè, near Moamba, fl. & fr. 21.ii.1948, *Torre* 7400 (C; COI; LD; LISC; LMU; M; MO; WAG).

Widespread in tropical and subtropical Africa and also in Madagascar, Arabia, Pakistan and India. In grasslands, floodplains, edges of swamps, riverbanks, roadsides and waste grounds, mainly on sandy or muddy soils; up to 1300 m.

4. **Heliotropium ciliatum** Kaplan in Ann. Transv. Mus. **12**: 187, t. 4 (1927). —Hilliard & Burtt in Notes Roy. Bot. Gard. Edin. **30**: 114 (1970). —J.H. Ross, Fl. Natal: 297 (1972). TAB. **26** fig. 3. Type from S. Africa (Transvaal).

    Tournefortia tuberculosa Cham. in Linnaea **4**: 467 (1829). —C.H. Wright in F.C. **4**, 2: 6 (1904). Type from S. Africa (Cape Prov.).

    Heliotropium tuberculosum (Cham.) Gürke in Engl. & Prantl, Pflanzenfam. **4**, 3a: 94 (1897). —Engl. in Sitz.-Ber. Königl. Preuss. Akad. Wiss. Berl. **52**: 875 (1907). —Rattray & Wild in Rhod. Agric. Journ. **52**: 499 (1955) nom. illegit., non Boiss. (1879).

    Heliotropium longiflorum sensu Baker & Wright in F.T.A. **4**, 2: 41 (1905) saltem pro parte quoad specim. *Kirk*.

    Heliotropium ciliatum var. *lanceolatum* Kaplan, loc. cit., t. 5 fig. A, a (1927). Type from S. Africa (Natal).

    Heliotropium pustulatum Kaplan, loc. cit., t. 5 fig. B, b (1927). Type from S. Africa (Orange Free State).

    Heliotropium gibbosum M. Friedr. in Mitt. Bot. Staatss. München **3**: 616 (1960). —Friedrich-Holzhammer in Merxm., Prodr. Fl. SW. Afr. **119**: 7 (1967). Type as for *Tournefortia tuberculosa*.

Erect rhizomatous perennial herb up to 80 cm. high. Stem usually branched from the base, somewhat swollen below the branches insertions, angular as for the branches with short appressed hairs. Leaves usually sparse; petiole 2–20(25) mm. long, with hair-covering like the branches; lamina 20–50 × 5–20 mm., lanceolate or ovate, with hard short appressed hairs above and on the nerves below, covered with the swollen bases of minute hairs below, rarely roughish on both surfaces, acute to obtuse and sometimes mucronate at apex, acute to obtuse, decurrent on the petiole and often asymmetrical at base, margins usually repand and undulate. Cymes lax, up to 19 cm. long when completely expanded, usually 2, rarely single, terminal; common peduncle 1–4 cm. long. Flowers sessile or subsessile. Calyx minute, callose-strigose or subglabrous outside, subglabrous or glabrous inside; lobes 0.5–1.0(2.0) mm. long, ovate to ovate-triangular, acute, callose-ciliate mainly at the base, thick. Corolla hypocrateriform, white, sparsely strigose outside on the upper part, otherwise glabrous outside and inside; tube 3.5–5.0 mm. long, slightly widened in the upper half; limb 3–5(6) mm. in diam.; lobes 1.2–2.2(2.5) mm. long, triangular-acuminate, usually with undulate and ± crenate margins, bent inwards in bud, spreading. Stamens inserted about the middle of the corolla tube; anthers 1.2–1.5 mm. long, narrowly oblong, emarginate at apex, obtuse to subcordate at base, subsessile. Ovary glabrous. Style 1.0–1.5(2.0) mm. long, filiform, glabrous; stigmatic ring not very well marked; sterile appendix 1.0–1.4 mm. long, narrowly conical, deeply 2-lobed, hirtellous. Fruits c. 4 × 5 × 4 mm., saddle-shaped at apex, tuberculate-crested, glabrous; nutlets connate in pairs, each pair 2-horned, 2-seeded, with the inner surface flat.

Botswana. N: Chobe Nat. Park, Savuti R., fl. & fr. 23.x.1972, *Pope, Biegel & Russell* 848 (K; SRGH). SW: 29 km. N. of Kang on road to Ghanzi, fl. & fr. 17.ii.1960, *De Winter* 7352 (K; PRE; SRGH). SE: Orapa, fl. & fr. 19.iii.1974, *Allen* 30 (PRE; SRGH). **Zimbabwe**. N: Darwin Distr., SE. corner of Chiswiti Reserve, fl. & fr. 22.i.1960, *Phipps* 2365 (K; SRGH). W: Matobo Distr., Champion Ranch, c. 8 km. WNW. of Shashi-Shashani confluence, fl. & fr. 8.v.1963, *Drummond* 8175 (K; SRGH). E: Sabi Valley Experimental Station, fl. & fr. i.1960, *Soane* 229 (SRGH). S: Beitbridge, fl. & fr. 16.ii.1955, *Exell, Mendonça & Wild* 450 (BM; LISC). **Mozambique**. T: Moatize, 3 km. from Necungas to Entroncamento, fl. & fr. 22.vi.1949, *Barbosa & Carvalho* 3230 (LMA). MS: Mungari, near Lupata, fl. & fr. ii.1859, *Kirk* s.n. (K). GI: Govuro, 3 km. from Banamana to Machaíla, fl. & fr. 19.iii.1974, *Correia & Marques* 4075 (LMU). M: Sábiè, Moamba, Mangulane, fl. & fr. 8.iv.1974, *Balsinhas* 2683 (LMA).

Also known from Angola, Namibia and S. Africa. In riverbanks, savannas with trees and shrubs and in dry woodlands, on sandy or rocky soils; up to 1050 m.

5. **Heliotropium zeylanicum** (Burm.f.) Lam., Encycl. Méth. Bot. **3**: 94 (1789). —Hiern, Cat. Afr. Pl. Welw. **1**: 720 (1898). —Schinz & Junod in Mém. Herb. Boiss. **10**: 60 (1900). —Baker & Wright in F.T.A. **4**, 2: 31 (1905). —Weimarck in Bot. Notis. **1940**: 62 (1940). TAB. **26** fig. 4. Type from India.

    Heliotropium curassavicum var. *zeylanicum* Burm. f., Fl. Ind.: 41, t. 16 fig. 2 (1768). Type as above.

    Tournefortia subulata Hochst. ex. A. DC. in DC., Prodr. **9**: 528 (1845). Syntypes from Senegal, Sudan and Ethiopia.

    Tournefortia stenoraca Klotzsch in Peters, Reise Mossamb. Bot.: 250 (1861). Type: Mozambique, Rios de Sena, *Peters* s.n. (B†, holotype).

    Heliotropium subulatum (Hochst. ex A.DC.) Vatke in Linnaea **43**: 316 (1882). —Friedrich-Holzhammer in Merxm., Prodr. Fl. SW. Afr., **119**: 9 (1967). —Taton in Fl. Congo, Rwanda et Burundi, Boraginaceae: 30, t. 4 (1971). —J.H. Ross, Fl. Natal: 297 (1972). —Munday & Forbes in Journ. S. Afr. Bot. **45**: 9 (1979). —Hall-Martin & Drumm. in Kirkia **12**: 177 (1980). Syntypes as for *T. subulata*.

Erect or procumbent perennial herb up to 60(90) cm. high, usually with a thick rootstock; stems somewhat woody to the base, hispid. Leaves sessile or subsessile, 20–65(120) × 2–14(20) mm., linear to linear-lanceolate, rarely lanceolate, pubescent-hispid above, glabrous or subglabrous below, however hispid on the nerves, acute at apex, cuneate at base, margins entire; often small orange sessile or pedicled glands are present mainly on the lower surface. Cymes lax, ebracteate, single or more often in pairs, up to 25(30) cm. long when completely expanded. Flowers sessile. Calyx 1.5–2.0 (2.5) mm. long, lobes narrowly ovate, acute, 3-nerved, 3-ribbed outside over the nerves, ribs usually undulate, sometimes only the middle one, rarely not perceptibly undulate, ciliate on the margins and on the midrib, sometimes glandulous outside, pubescent to the apex inside, stellate-spreading after the nutlets are shed. Corolla yellow or greenish; tube 3.5–4.5(5.5) mm. long, widened to the upper half, pubescent or strigose outside by strips corresponding the lobes; corolla lobes 1.5–3.0 mm. long, caudate-acuminate, spreading, with the tip bent back inwards, at least in bud. Stamens inserted between $\frac{1}{2}$ and $\frac{2}{3}$ from the base on the corolla tube; anthers 1.0–1.5 mm. long, lanceolate to oblong, shortly bifid at apex, subsessile. Ovary ovoid, glabrous. Style 1.0–2.0 mm. long, glabrous; stigmatic ring small; sterile appendix 0.8–1.3 mm. long, narrowly cone-shaped, brush-like. Fruits 1.7–2.3 × 2.0–2.5 mm., subglobulous, glabrous, rugose, usually bright brown; nutlets free, with the inner surfaces ± convex, irregular.

Botswana. N: Dikgakana, Mogohilo R. area, fl. & fr. 16.v.1976, *Smith* 1740 (K; SRGH). **Zambia**. B: Sesheke, fl. & fr. i.1925, *Borle* in N.H. 58179 (PRE). N: Mbala, Kasikalawi, W. of Mpulungu, fl. & fr. 10.iv.1961, *Phipps & Vesey-FitzGerald* 3040 (K; LISC; SRGH). C: Katondwe, fl. & fr. 4.xii.1969, *Mutimushi* 3869 (SRGH). E: Chipata, Kakumbi, fl. & fr. 21.i.1970, *Astle* 5751 (K; SRGH). S: Mumbwa-Namwala Rd., 1.6 km. E. of Lukomezi R., fl. & fr. 15.viii.1963, *van Rensburg* 2456 (K; SRGH). **Zimbabwe**. N: Guruve, near Mwanzamtanda R., fl. & fr. 28.i.1966, *Müller* 238 (K; LISC; SRGH). W: Victoria Falls, fl. & fr. vii.1920, *Sim* 19155 (PRE). E: Chimanimani, near Chitsa's Village, fl. & fr. 11.vi.1950, *Chase* 2405 (BM; SRGH). S: Mwenezi, near Palfrey's Store, fl. & fr. 29.iv.1961, *Drummond & Rutherford-Smith* 7601 (K; LISC). **Malawi**. N: S. Rukuru R. Gorge, fl. & fr. 8.vii.1974, *Pawek* 8810 (K; SRGH). C: near Kasungu Hill, fl. & fr. 14.i.1959, *Robson & Jackson* 1147 (BM; K; LISC; SRGH). S: Mangochi, E. of Lake Malombe, fl. & fr. 16.ix.1963, *Salubeni* 107 (SRGH). **Mozambique**. T: Cahobra (Cahora) Bassa, serra de Songa, fl. 30.xii.1965, *Torre & Correia* 13926 (C; J; LISC; LMU; MO). MS: Búzi, Chibabava, fl. & fr. 30.xi.1906, *Swynnerton* 1941 (BM; K). GI: Caniçado, Guijá, fl. & fr. 3. vii.1947, *Pedro & Pedŕogão* 1241 (COI; K; LMA; SRGH). M: Inhaca Isl., fl. & fr. 19.vii.1958, *Mogg* 28100 (BM; K; LMA; LMU; SRGH).

Widespread throughout tropical Africa and in Arabia, Pakistan and India. In dry woodlands and savannas, open grounds, on sandy riverbanks and on edges of rivers and lakes; up to 2000 m.

6. **Heliotropium giessii** Friedr. in Mitt. Bot. Staatss. München **3**: 616 (1960). —Friedrich-Holzhammer in Merxm., Prodr. Fl. SW. Afr. **119**: 7 (1967). TAB. **26** fig. 5. Syntypes from Namibia.

    Heliotropium erectum Holzhammer in Mitt. Bot. Staatss. München **1**: 338 (1953) nom. illegit., non Lam. (1778) nec Vell. (1825).

Erect perennial herb up to 50(100) cm. high, woody at the base, strong; stem and branches subterete, clothed with short white retrorse hairs and with some longer antrorse ones, rarely all antrorse. Leaves alternate; petiole 0–8(15) mm. long with hair-covering like the branches; lamina 30–120 × 8–25 mm., lanceolate to oblong, hairy or rough above, hairy to tomentose below, densely covered with the swollen bases of the hairs, acute to rounded at apex, acute to obtuse at base, often asymmetrical, margins entire or repand, secondary nerves 4–5 on each side of the midrib. Cymes dense usually short, in groups of 2–4(5) on each terminal peduncle. Flowers sessile or subsessile. Calyx villous outside and inside; lobes 3.0–3.5 × 0.5–0.8 mm., narrowly ovate to narrowly oblong, acute to rounded at apex, erect. Corolla white; tube 4–6 mm. long, slightly swollen over the anthers, clothed with short retrorse hairs outside, glabrous inside, with 5 longitudinal folds on the second third inside, on the line of the corolla lobes; limb 3–4 mm. in diam., spreading, subglabrous, lobes 1.2–1.5 × 0.8–1.3 mm., oblong or oblong-obovate, rounded at apex and with undulate margins. Stamens inserted at $\frac{2}{3}$ from base on the corolla tube; anthers 1.2–1.5 mm. long, lanceolate, apiculate, apex reaching the mouth, subsessile. Ovary ovoid, glabrous. Style 2–3 mm. long, swollen to the apex, with very short retrorse hairs on the upper half; stigmatic ring c. 0.5 mm. in diam.; sterile appendix 0.6–0.8 mm. long, conical, bifid, roughish or hirtellous. Fruits c. 2.5 × 3.0 mm., subglobulous, 4-lobed, pustulate, enclosed in the calyx; nutlets free.

Botswana. N: Ngamiland, Nxai Pan, fl. & fr. 22.ii.1966, *Drummond* 8848 (K; SRGH). SW: Central Kalahari Game Res., Deception Pan, fl. & fr. 1.iv.1975, *Owens* 78 (K; SRGH). SE: Takatokwane Pan, fl. & fr. 17.ii.1960, *Wild* 4989 (K; SRGH). **Zimbabwe**. W: Insiza, Filabusi, fl. & fr. 8.ii.1974, *Mavi* 1524 (K; SRGH). S: Beitbridge, Shashe-Limpopo confluence, fl. & fr. 22.iii.1959, *Drummond* 5946 (K; LISC; SRGH).
Also recorded from S. Kenya, N. Tanzania, Namibia and S. Africa (Transvaal). On bare soils of the pans and in open short woodlands; 450–1200 m.

7. **Heliotropium lineare** (A.DC.) Gürke in Engl. & Prantl, Pflanzenfam. **4**, 3a: 96 (1897). —C.H. Wright in F.C. **4**, 2: 9 (1904). —Bremek. & Oberm. in Ann. Transv. Mus. **16**: 430 (1935). —Friedrich-Holzhammer in Merxm., Prodr. Fl. SW. Afr. **119**: 8 (1967). TAB. **26** fig. 6. Syntypes from S. Africa (Cape Prov.).
    Tournefortia linearis E. Meyer ex Drège in Flora **26**, 2, Beigabe: 57, 226 (1843) nom. tantum.
    Heliophytum lineare A.DC. in DC., Prodr. **9**: 555 (1845). Syntypes as above.

Erect perennial herb up to 45(60) cm. high, with a thin rootstock and usually branched from the base; stem and branches angular, ± densely strigose with 2-armed trichomes. Leaves 18–60(75) × 1–4(8) mm., linear basal ones usually narrowly elliptic, rough on both surfaces, strigose with 2-armed trichomes on the midrib and also sometimes on the secondary nerves below, acute to obtuse at apex, gradually narrowed into a short petiole at base, with entire revolute margins. Cymes short, up to 9 cm. long when completely expanded, lax, ebracteate, often 2–4 together on short terminal peduncles. Flowers sessile or subsessile. Calyx subglabrous outside, glabrous or subglabrous inside; lobes 1.0–1.8 mm. long, narrowly ovate to narrowly triangular, subequal, acute to obtuse, erect. Corolla narrowly infundibuliform, white, cream or yellow, clothed with scale-like 2-armed trichomes outside, glabrous inside; tube (3)4–5 mm. long; lobes 1.2–2.0 mm. long, triangular-acuminate and with undulate margins, subequal. Stamens inserted about the middle of the corolla tube; anthers 1.0–1.3 mm. long, narrowly oblong, apiculate, subsessile. Ovary ovoid, glabrous. Style 1.0–1.6 mm. long, terete, glabrous; stigmatic ring c. 0.5 mm. in diameter, well-marked; sterile appendix c. 1 mm. long, conical, bi or 4-fid, glabrous. Fruits c. 2.0 × 3.5 × 2.0 mm., transversely ellipsoid, deeply depressed laterally, granulate-verrucose; nutlets free, usually narrowly winged, with the inner faces ± flat; each nutlet includes 2 cavities, the distal one fertile, the proximal one empty and usually smaller than the other.

Botswana. SW: Kubi, Farm 102, fl. & fr. 21.ii.1970, *Brown* 8724 (K; PRE; SRGH). SE: Takatokwane Pan, fl. & fr. 17.ii.1960, *Wild* 4987 (K; SRGH).
Also in Namibia and S. Africa. In muddy pans and sometimes on wooded stony habitats; 900–1350 m.
Probably not specifically distinct from *H. longiflorum* (A.DC.) Jaub. & Spach (with syn. *H. undulatifolium* Turrill, *H. engleri* Vaupel, *H. somalense* Vatke) from east Africa, but I think that this matter needs further study.

8. **Heliotropium elongatum** (Lehm.) I.M. Johnston in Contr. Gray Herb., part 8l: 18 (1928). — Hilliard & Burtt in Notes Roy. Bot. Gard. Edin. **34**: 75 (1975). TAB. **26** fig. 7. Type from Brazil.
    Tiaridium elongatum Lehm., Pl. Asper.: 16 (1818); in Icon. Rar. Pl. Asper. t. 6 (1821). —Chamisso in Linnaea **4**: 452, t. 5 fig. 1 (1829). Type as above.

Erect annual herb up to 90 cm. (or more) high, little branched; stem and branches bristly on the young parts, glabrescent. Leaves alternate or subopposite; petiole up to 20 mm. long or probably longer, bristly; lamina 40–50 × 25–35 mm. on specimens seen but probably reaching magnitude as for *H. indicum*, ovate to broadly ovate, strigose by areolas and with scattered bristles above, roughish to pubescent and bristly on nerves and veins below, acute at apex, obtuse at base and abruptly decurrent on the petiole, with slightly undulate to crisped margins. Cymes up to 17 cm. long in specimens seen, single, scorpioid, terminal. Flowers sessile. Calyx divided to $\frac{2}{3}$-$\frac{3}{4}$, bristly outside, strigose inside; lobes 1.4–1.7 mm. long, narrowly triangular, acute, subequal or unequal. Corolla hypocrateriform, bristly outside, glabrous inside, purple, pale blue or mauve, with yellow throat; tube 3.5–5.0 mm. long, plicate at the throat; limb c. 3 mm. in diam., subentire or shallowly divided; lobes up to 0.3 × 1.0 mm., depressed-ovate, usually obtuse or rounded. Stamens inserted at c. 1.5 mm. from the base on the corolla tube; anthers c. 1 mm. long, linear, rounded on both ends, sessile. Ovary ovoid, glabrous. Style 0.5 mm. long, glabrous; stigmatic ring well marked; sterile appendix hemispheric, papillous. Fruit c. 4.0 × 3.5 × 3.0 mm., bilobed at apex and mitre-shaped, the lobes little divergent, ribbed, brownish, glabrous; nutlets late free, ovoid, beaked, with the out surface 2-4-ribbed and the inner one slightly tuberculate, in cross section showing only one seed-bearing cavity.

Malawi. S: Blantyre Distr., Mpatamanga Gorge, fl. & fr. 5.ii.1979, *Blackmore & Patel* 287 (K).

Native of South America and well established in Java. In the Flora Zambesiaca area it occurs on sandy or muddy soils of Shire R. banks.

Superficially similar to *H. indicum* L. from which it differs mainly by its fruits. In *H. elongatum* the fruit lobes are subparallel or slightly divergent, and in cross-section each nutlet has a single seed-bearing cavity. In *H. indicum* the fruit-lobes are strongly divergent, and in cross-section each nutlet has one seed-bearing and one larger empty cavity. Immature plants are sometimes difficult to identify.

H. elongatum seems to be a recent introduction into the Flora Zambesiaca area. Indeed, the first collection known is dated 1967 (*Hilliard & Burtt* 4160) from the upper Shire R., the third and latest is dated 1979 (*Blackmore & Patel* 287) and has written on the label "Common. Amongst rocks on sandy soil at side of Shire River" which seems to show that this species became naturalized.

9. **Heliotropium indicum** L., Sp. Pl.: 130 (1753). —Hiern, Cat. Afr. Pl. Welw. **1**: 719 (1898). —Baker & Wright in F.T.A. **4**, 2: 32 (1905). —Schinz in Denkschr. Akad. Wiss. Wien, Math-Nat. Kl. **78**: 436 (1905). —Brenan in Mem. N.Y. Bot. Gard. **9**: 6 (1954). —Friedrich-Holzhammer in Merxm., Prodr. Fl. SW. Afr. **119**: 8 (1967). —Binns, H.C.L.M.: 24 (1968). —Hilliard & Burtt in Notes Roy. Bot. Gard. Edin. **30**: 116 (1970). —Taton in Fl. Congo, Rwanda et Burundi, Boraginaceae: 29 (1971). —J.H. Ross, Fl. Natal: 297 (1972). TAB. **26** fig. 8. Type probably from Sri Lanka.

Tiaridium indicum (L.) Lehm., Pl. Asper.: 14 (1818). Type as above.

Heliophytum indicum (L.) DC., Prodr. **9**: 556 (1845). Type as above.

Erect annual or a few times short-lived perennial herb up to 150 cm. high. Stem quadrangular, thick, empty or pithed, woody to the base, variably hairy. Leaves alternate or subopposite; petiole 20–70(110) mm. long, pubescent, with scattered bristles; lamina 40–170 × 30–130 mm., ovate to broadly ovate or triangular-ovate, strigose by areolae and with scattered bristles, roughish or only punctate above, generally pubescent-villous, rarely tomentose or punctate and with some scattered bristles mainly on the nerves and veins below, acute at the apex, obtuse to subcordate at the base and abruptly decurrent on the petiole, often with crisped margins; secondary nerves 4–8 on each side of the midrib. Cymes long and flexuous, up to 36 cm. long when completely expanded, usually shorter, ± dense, single or sometimes in pairs, on a short terminal peduncle. Flowers sessile. Calyx lobes 1.5–3.5 mm. long, narrowly elliptic to subulate, unequal, sparsely hispid. Corolla 3.5–5.0 mm. long, hypocrateriform, pubescent outside, glabrous inside, white or rarely mauve or blue; tube 3.0–4.5 mm. long, usually longer than the calyx; lobes c. 0.6 × 1.0 mm., depressed-ovate, rounded at apex. Stamens inserted at 0.8–1.0 mm. from the base of the corolla tube; anthers 0.5–0.9 mm. long, narrowly ovate, cordate at the base, subsessile. Ovary with 4 fleshy crests, glabrous. Style 0.5 mm. long, terete, glabrous; sterile appendix hemispheric. Fruit deeply bilobed, mitre-shaped with the lobes strongly divergent, glabrous or puberulous, first divided into 2, each one c. 4 × 4 × 2.5 mm., later split into 4 nutlets, these angular, rostrate, with the surface 3-ribbed; each nutlet with 2 cavities, the distal one fertile and the proximal one empty and larger than the other.

Caprivi Strip. Katima Mulilo area, c. 72 km. from Katima to Lisikili, fl. & fr. 24.xii.1958, *Killick & Leistner* 3080 (K). **Botswana**. N: Chobe Nat. Park, fl. & fr. 19.xi.1969, *Mahundu* CNP/2/19 (SRGH). **Zambia**. B: Mongu, fl. & fr. 25.xi.1964, *van Rensburg* 3011 (K; SRGH). C: Luangwa Valley, Mfuwe, fl. & fr. 26.xi.1965, *Astle* 4126 (K). C: Luangwa (Feira), fl. & fr. 27.ix.1962, *Angus* 3342 (K). E: Nsefu Game Camp, fl. & fr. 15.x.1958, *Robson & Angus* 114 (BM; K; LISC; SRGH). S: Lusitu, fl. & fr. 19.v.1960, *Fanshawe* 5673 (K; SRGH). **Zimbabwe**. N: Hurungwe, Sanyati-Chiroti junction, fl. & fr.

21.xi.1953, *Wild* 4224 (K; LISC; SRGH). E: Chipinge Distr., Sabi Valley Experimental Station, fl. & fr. xi.1959, *Soane* 159 (K; LISC; SRGH). S: Mwenezi, near Malipate, fl. & fr. 2.v.1961, *Drummond & Rutherford-Smith* 7682 (K; LISC; SRGH). **Malawi.** C: Ntchisi Distr., Bua R. on Kasungu-Nkhota Kota Rd., fl. 20.vi.1970, *Brummitt* 11582 (K). S: Chiromo, Lower Shire Valley Belt, fl. & fr. 29.ii.1968, *Banda* 988 (K; SRGH). **Mozambique.** N: Montepuez, R. M'salo (Messalo), fl. & fr. 28.viii.1948, *Pedro & Pedrogão* 4973 (LMA). Z: Morrumbala, Vila Bocage, R. Shire, fl. & fr. 13.xii.1971, *Pope & Müller* 600 (LISC; SRGH). T: Cahobra (Cahora) Bassa, mouth of R. Mecangádzi, fl. & fr. 2.xi.1973, *Correia, Marques & Diniz* 3675 (LMU). MS: Gorongosa, Nat. Game Res., fl. & fr. 6.xi.1963, *Torre & Paiva* 9075 (C; LD; LISC; LMU; MO). GI: Gaza, near Mabalane, fl. & fr. 4.vi.1959, *Barbosa & Lemos* 8604 (COI; K; LISC; LMA). M: Matutuíne (Maputo), Bela Vista, Tinonganine, fl. & fr. 12.xii.1967, *Gomes e Sousa & Balsinhas* 5013 (LMA).

Pantropical and also in the south of the United States of America. On sandy or muddy riverbanks, edges of rivers and lagoons, in floodplains, old gardens near rivers and pastures, always near water; up to 1100 m.

See note under *Heliotropium elongatum* (Lehm.) I.M. Johnston.

10. **Heliotropium harareense** E. Martins in Garçia de Orta, Sér. Bot. **9**: 74 (1988). TAB. **26** fig. 9.
    Type: Zimbabwe, Harare, Mabelreign, *Greatrex* in GHS 26562 (K, holotype).

Erect annual herb up to 65 cm. high, often much smaller; stem often unbranched, sometimes few-branched, rarely branched from near the ground, ± densely clothed with short spreading or slightly retrorse bristles. Lower leaves opposite or subopposite, the upper ones alternate, 30–120 × 10–50 mm., lanceolate to oblong-oblanceolate or sometimes rhombic, rough above, densely clothed with minute tubercle-based hairs to subglabrous but shortly bristly on the nerves below, acute to obtuse at apex, cuneate at base and gradually merged into a petiole, membranous to papery, with 5–6 slightly raised nerves on each side of the midrib. Cymes up to 12 cm. long when full expanded, 2, rarely 3 together on each terminal peduncle. Flowers sessile. Calyx 1.5–2.5 mm. long, with some short and scattered bristles; lobes narrowly triangular to linear-subulate, acute. Corolla white; tube 2.0–2.5 mm. long, slightly strigose outside, glabrous inside; lobes 0.8–1.5 mm. long, circular to oblong-obovate, rounded at apex, with slightly undulate margins, unequal, spreading. Stamens inserted below the middle of the corolla tube; anthers 0.8–11.2 mm. long, lanceolate, acute, glabrous, often glaucous, subsessile. Ovary ovoid, glabrous. Style 0.4–0.5 mm. long, terete, glabrous; sterile appendix 0.4–0.6 mm. long, conical, minutely papillose with a well-marked stigmatic ring at base. Fruits c.3.0 × 5.0 × 3.5 mm., much depressed, brown, shining; nutlets connate in pairs, each pair with a wide groove on the back, often strongly asymmetrical by abortion.

Zambia. C: Luangwa (Feira), fl. & fr. immat. 16.xii.1970, *Fanshawe* 11027 (SRGH). **Zimbabwe.** N: Hurungwe Distr., Mensa pan, 17.5 km. ESE. of Chirundu bridge, fl. & fr. immat., 29.i.1958, *Drummond* 5318 (SRGH). W: Bulawayo Distr., fl. & fr. immat. xii.1901, *Eyles* 20 (SRGH). C: Norton Distr., fl. & fr. immat. 2.i.1967, *Biegel* 1626 (K; SRGH). **Malawi.** N: Lupembe, 24 km. S. of Karonga, fl. & fr. 11.ii.1953, *Williamson* 147 (BM).

Also known from Tanzania. In woodlands, wet veldt, and termite mounds, also as weed in cultivated lands, very often on black soils; 450–1370 m.

11. **Heliotropium steudneri** Vatke in Öst. Bot. Zeitschr. **25**: 167 (1875) emend. in Linnaea **43**: 320 (1882). —Baker & Wright in F.T.A. **4**, 2: 42 (1905). —Hilliard & Burtt in Notes Roy. Bot. Gard. Edin. **30**: 116 (1970). —J.H. Ross, Fl. Natal: 297 (1972). TAB. **26** fig. 10. Type from Ethiopia.
    Heliotropium eduardii Martelli, Flor. Bogos: 59 (1886). —Baker & Wright, tom. cit.: 43 (1905). Type from Ethiopia.
    Heliotropium longiflorum sensu Hiern, Cat. Afr. Pl. Welw. **1**: 720 (1898). —Baker & Wright, tom. cit.: 41 (1905) pro parte quoad specim. Angol.
    Heliotropium nelsonii C.H. Wright in F.C. **4**, 2: 9 (1904). —Friedrich-Holzhammer in Merxm., Prodr. Fl. SW. Afr. **119**: 8 (1967). Syntypes from S. Africa.
    Heliotropium dissimile N.E. Br. ex Baker & Wright, tom. cit.: 42 (1905); in Kew Bull. **1909**: 122 (1909). Types: Botswana, Ngamiland, Kgwebe, *F.D. Lugard & E.J. Lugard* 139 (K, syntype); and Kgwebe Hills, *Mrs Lugard* 77 (K, syntype; LISC, icon).
    Heliotropium rogersii Kaplan in Ann. Trans. Mus. **12**: 188, t. 7 (1927). Type from S. Africa (Transvaal).

Erect or procumbent perennial herb 60–90 cm. high, woody to the base, sometimes with a thick rootstock; stems and branches ± densely clothed with short appressed hairs with a few longer spreading ones. Leaves alternate, sessile or with petiole up to 6 mm. long; lamina 20–60(80) × 4–15 mm., lanceolate, pubescent to rough above, pubescent to subglabrous below, acute to obtuse at apex, cuneate at base and sometimes decurrent on the petiole, margins slightly revolute and sometimes undulate to crisp and with 3–5(6) secondary nerves on each side of the midrib. Cymes up to 12(18) cm. long when full expanded, single or in pairs, terminal. Flowers sessile or subsessile. Calyx 2.5–3.5(4.5) mm. long, hairy outside, hairy to glabrous inside; lobes narrowly triangular to linear, obtuse, stellate-spreading after the nutlets are shed. Corolla quite white or with yellow throat; tube 3.5–5.5 mm. long, appressed hairy outside, glabrous inside; lobes 1.3–2.0 × 1.1–1.8 mm. oblong to subquadrate, obtuse to truncate, usually with very undulate margins, unequal, spreading. Stamens inserted at the middle of the corolla tube; anthers 1.4–1.8 mm. long, oblong, apiculate, glabrous, often glaucous, subsessile. Ovary ovoid-conical, glabrous. Style 1.0–1.5 mm. long, terete, glabrous; stigmatic ring 0.5–0.7 mm. in diam.; sterile appendix 0.7–1.0 mm. long, as broad at the base as the stigmatic ring, papillose. Fruits 3–4 mm. long, coarsely obpyriform and depressed between the pairs of nutlets or more often strongly asymmetrical by abortion of 1–3 seeds, smooth, glabrous; nutlets connate in pairs, each pair emarginate at apex.

Botswana. N: Makalamabedi A.I. Camp, fl. & fr. 27.xi.1974, *Smith* 1203 (K; SRGH). SW: Ghanzi, Farm 48, fl. & fr. 25.ii.1969, *De Hoogh* 83 (K). SE: 51 km. W. of Kanye, fl. & fr. 18.i.1960, *Leach & Noel* 199 (K; SRGH). **Zimbabwe**. W: Bulawayo, Lochview, fl. & fr. 25.x.1975, *Cross* 249 (K; SRGH). E: Odzi R., Sun Valley Farm, fl. 18.xii.1954, *Chase* 5367 (BM; K; SRGH). S: Mwenezi, near Palfrey's Store, fl. & fr. 29.iv.1961, *Drummond & Rutherford-Smith* 7600 (K; LISC; SRGH). **Mozambique**. GI: near Combomune, fl. 12.v.1948, *Torre* 7776 (COI; LISC; LMU). M: Namaacha, between Goba and Changalane, fl. 28.ii.1952, *Myre & Carvalho* 1177 (K; LISC; LMA; SRGH).
Also known from east Africa, south-western Angola, Namibia and S. Africa. In open grounds, grasslands and in dry bushes on dry, poor and often sandy soils; 100–1350 m.

12. **Heliotropium supinum** L., Sp. Pl.: 130 (1753). —DC., Prodr. **9**: 533 (1845). —Gürke in Engl., Pflanzenw. Ost-Afr. **C**: 336 (1895). —Hiern, Cat. Afr. Pl. Welw. **1**: 717 (1898). —C.H. Wright in F.C. **4**, 2: 8 (1904). —Baker & Wright in F.T.A. **4**, 2: 37 (1905). —Friedrich-Holzhammer in Merxm., Prodr. Fl. SW. Afr. **119**: 9 (1967). TAB. **26** fig. 11. Lectotype from Montpellier (France).

Prostrate or decumbent annual herb usually with 4 main stems rising in a cross; stems up to 35 cm. long, herbaceous, laterally compressed, ± densely clothed with retrorse appressed hairs, mixed with longer spreading ones. Leaves alternate however subopposite to the base; petiole 3–15 mm. long (up to 27 mm. on the basal leaves); lamina 10–35 × 7–19 mm., broadly ovate to broadly obovate, rarely trullate or circular, densely appressed-pubescent or strigose above, pubescent or tomentose and bristly on the nerves below, obtuse to truncate at apex, acute to obtuse and often very asymmetrical at base, the nerves usually deepened above, prominent below. Cymes short, often 1, rarely 2–3 on each terminal peduncle. Flowers sessile or subsessile. Calyx 2.0–2.8 mm. long, enlarged up to 4.5(6) mm. in fruit, bristly outside, appressed hairy inside; lobes cohering up to near apex, obtuse. Corolla as long as or but little longer than the calyx, persisting over the fruit; tube greenish, densely clothed with retrorse appressed hairs outside, glabrous inside; limb white, divided into 5 minute oblong subequal lobes. Anthers c. 1 mm. long, linear, subsessile. Ovary globose, glabrous. Style c. 0.5 mm. long with short retrorse hairs, sometimes subglabrous; sterile appendix c. 0.5 mm. long, conical, shortly hairy at apex. Fruits up to 4 mm. long, ovoid, plano-convex or excavated by abortion of 2–3 nutlets, glabrous, enclosed in the calyx; nutlets often 2, rarely 1, very rarely 3 or 4 with the outer and the inner surfaces granulate-tuberculate, brown with a lighter well developed edge; rarely the fruits wholly covered with a greyish corky layer concealing the ornamentation.

Botswana. N: Linyanti R., c. 3 km. E. of Shaile Camp, fl. & fr. 28.x.1972, *Pope, Biegel & Russell* 886 (COI; K; SRGH). SE: 8 km. E. of Letlhakane (Lothlekane), fl. & fr. 23.iii.1965, *Wild & Drummond* 7253 (K; LISC; SRGH). **Zambia**. B: Masese, fl. & fr. 17.vi.1962, *Fanshawe* 6887 (SRGH). C: Mpika Distr., Mushilashi-Luangwa River confluence, fl. 4.v.1965, *Mitchell* 2793 (K; SRGH). S: Mazabuka Distr., Kafue Flats, fl. & fr. 20.vii.1941, *Greenway* 6233 (K). **Zimbabwe**. N: Hurungwe Distr., Nyamachera R., fl. & fr. viii.1970, *Guy* 1072 (K; LISC; SRGH). W: Hwange Nat. Park, Mosuma Dam, fl. & fr. 7.xii.1966, *Clark* 464 (K; SRGH). C: Umsweswe, fl. 20.iv.1921, *Borle* 177 (K; PRE). S: Chiredzi, Colopo R., Chipinda Pools, fl. & fr. 31.v.1970, *Sherry* 155/70 (K; SRGH). **Mozambique**. T: Tete, Sisitso, R. Zambezi, fl. & fr. 19.vii.1950, *Chase* 2679 (BM; K; SRGH). M: Magude, 6 km. from Mapulanguene to Massingir, fl. 26.vi.1969, *Correia & Marques* 856 (LMU).
Also in subtropical Africa, southern Europe, Syria, Arabia, Iraq and India. On heavy black or brown soils in dried-up water holes and pools; 250–1350 m.

13. **Heliotropium curassavicum** L., Sp. Pl.: 130 (1753). —DC., Prodr. **9**: 538 (1845). —Hiern, Cat. Afr. Pl. Welw. **1**: 718 (1898). —C.H. Wright in F.C. **4**, 2: 7 (1904) "currassavicum". —Baker & Wright

in F.T.A. **4**, 2: 35 (1905). —Friedrich-Holzhammer in Merxm., Prodr. Fl. SW. Afr. **119**: 7 (1967). —J.H. Ross, Fl. Natal: 297 (1972). TAB **26** fig. 12. Type from tropical America (Curaçao).

Prostrate or decumbent annual or perennial fleshy herb, much branched, glaucous, quite glabrous. Leaves alternate or supposite, 15–30(45) × 2–6 mm., narrowly oblong to narrowly obovate, rounded at apex, narrowed gradually to the base into a short petiole, entire, fleshy, slightly verrucous. Cymes moderately dense, ebracteate, slightly scorpioid, up to 8 cm. long when completely expanded, single or 2, rarely to 4 on each short common peduncle, terminal. Flowers subsessile. Calyx c. 2 mm. long, irregularly divided; tube none to as long as ⅓ of the calyx; lobes usually subequal, ovate to narrowly triangular, apex subacute to rounded, slightly inflexed. Corolla usually overlapping the calyx a little, salver-shaped, white; tube 1.5–2.0 mm. long; lobes usually unequal, the largest one up to 1.0 mm. long, triangular-ovate to oblong, rounded at apex. Stamens inserted at $\frac{1}{3}-\frac{1}{2}$ of the corolla tube; anthers c. 0.8 mm. long, narrowly ovoid, apiculate, subcordate at base, subsessile. Ovary ovoid. Style very short; stigma umbrella shaped without sterile appendix. Fruits c. 2 mm. long, ± globulose; nutlets ovoid, free at maturity, with the outer surface rugose, the inner one with a ± circular cavity or a shell-shaped depression.

Botswana. N: Boteti (Botletle) R., fl. & fr. 20.ii.1980, *P.A. Smith* 3118 (K). **Mozambique**. MS: Sofala, on the road from Nova Mambone to Quicuaxa, fl. & fr. 2.ix.1942, *Mendonça* 120 (COI; LISC; LMU; WAG). GI: Chibuto, Baixo Changane, fl. & fr. 23.viii.1963, *Macêdo & Macúacua* 1131 (K; LMA). M: Maputo (Lourenço Marques), Costa do Sol, fl. & fr. 5.xi.1963, *Balsinhas* 662 (K; LISC; LMA).

Native in tropical America, however, at present widely introduced and naturalized in S. Europe and in tropical and subtropical regions of the Old World and Australasia. A typical halophyte.

7. TRICHODESMA L.

By R. K. Brummitt

Trichodesma R. Br., Prodr. Fl. Nov. Holl.: 496 (1810) nom. conserv.
Borraginoides Boehmer in Ludwig, Def. Gen. Pl.: 18 (1760).
Friedrichsthalia Fenzl in Endlicher & Fenzl, Nov. Stirp. Dec.: 53 (1839).
Leiocarya Hochst. in Flora **27**: 30 (1844).
Boraginella Kuntze, Rev. Gen. Pl. **2**: 435 (1891).

Annual or perennial herbs to small shrubs, often with conspicuously tuberculate hairs. Lower leaves usually opposite or sub-opposite, upper ones usually alternate. Inflorescence either a bracteate one-sided cyme or a branching system of ebracteate very lax cymes. Flowers (calyx, corolla and stamens) 5-merous or occasionally some flowers in *T. zeylanicum* 6-merous. Sepals free to the base or loosely adherent along lower margins, usually ovate to lanceolate and rounded to cordate at the base, strongly accrescent in fruit. Corolla campanulate to funnel-shaped, with ovate-acuminate to broadly triangular lobes which are often spreading or reflexed, without scales at the throat. Stamens sessile or subsessile, the filament absent or less than 1 mm. long and broader than long, the anthers oblong and inserted on the inner side of the connectives which are prolonged above the anthers, the whole androecium converging above to form a cone with the connective prolongations twisted together at the apex, the connectives with long shaggy hairs on their dorsal surface. Ovary 4-lobed, with straight gynobasic style, the stigma scarcely differentiated. Fruit either of four smooth or variously ornamented nutlets attached to a pyramidal gynobase or (by abortion) of a single sessile nutlet (see note on sections below).

A genus of perhaps 40 species in tropics and sub-tropics of the Old World. It includes a remarkably wide range of fruit types, which elsewhere in the *Boraginaceae* might well have been the basis for recognition of several genera. However, the floral and pollen characters are relatively constant and it is preferable to recognise only a single genus. Brand, Pflanzenreich 78, [IV, 252]: 19–44 (1921) divided it into six sections based on fruit characters; of the species in our area *sp. 1* falls into what must now be called sect. *Trichodesma* (sect. *Leiocaryum* (A.DC.) Brand) with four smooth nutlets, *sp. 2* falls into sect. *Acanthocaryum* Brand with four strongly glochidiate nutlets, and *spp. 3* and *4* into sect. *Friedrichsthalia* (Fenzl) A.DC. non Brand (sect. *Trichocaryum* Brand nom. illegit.) with a single sessile tomentose nutlet.

1. Pedicels and inflorescence axes densely clothed with flexuous spreading hairs 1–2 mm. long; corolla less than 10 mm. long; nutlets 4, smooth   -   -   -   -   -   1. *zeylanicum*

– Pedicels and inflorescence axes glabrous or with appressed or stiff and often hooked setae;
   corolla more than 10 mm. long; nutlets 1 or 4, tomentose or glochidiate   -   -   -   2.
2. Leaves 0.1–0.6(0.9) cm. broad, linear to linear-elliptic; sepals 2.5–4 mm.
   broad   -   -   -   -   -   -   -   -   -   -   -   2. *angustifolium*
– Leaves (0.2)0.5–3(3.6) cm. broad, linear-lanceolate to ovate or elliptic; sepals (3)3.5–8(9) mm.
   broad   -   -   -   -   -   -   -   -   -   -   -   -   3.
3. Stems glabrous or with few setose hairs on lower internodes; sepals in flower mostly 3.5–6 mm.
   broad; corolla white or pinkish, with lobes not markedly spreading or reflexed and rounded
   with a cuspidate apex   -   -   -   -   -   -   -   -   -   3. *physaloides*
– Stems setose-pubescent, occasionally only sparsely so; sepals in flower 5–8 mm. broad; corolla
   blue or sometimes pinkish, with lobes spreading or reflexed and triangular with a long acute
   apex   -   -   -   -   -   -   -   -   -   -   4. *ambacense* subsp. *hockii*

1. **Trichodesma zeylanicum** (Burm. f.) R. Br., Prodr. Fl. Nov. Holl.: 496 (1810). —Klotzsch in Peters,
   Reise Mossamb., Bot.: 255 (1861). —Baker & Wright in Thistelton-Dyer, F.T.A. **4**, 2: 51 (1905).
   —Schinz, Pl. Menyharth.: 70 (1905). —Eyles in Trans. Roy. Soc. S. Afr. **5**: 457 (1916). —Brand in
   Engl., Pflanzenr. **78**, [IV, 252]: 40 (1921). —Gomes e Sousa in Bol. Soc. Estud. Colon. Moçamb.
   26: 49 (1935); op. cit. 32: 87 (1936). —Brenan in Mem. New York Bot. Gard. **9**: 6 (1954).
   —Chapman, Veg. Mlanje Mt.: 33 (1962). —Binns, H.C.L.M.: 24 (1968). —Richards & Morony,
   Check List Fl. Mbala Distr.: 208 (1969). —Taton in Fl. Congo, Rwanda et Burundi,
   Boraginaceae: 37 (1971). —Biegel & Mavi in Wild, Rhod. Bot. Dict. ed. 2: 260 (1972). —Ross, Fl.
   Natal: 297 (1973). —Agnew, Upl. Kenya Fl.: 520 (1974). —Hall-Martin & Drummond in Kirkia
   12: 177 (1980). Type from Sri Lanka.
   Borago zeylanica Burm. f., Fl. Ind.: 41, t. 14(2) (1768). —Linn. f., Mantissa **2**: 202 (1771). Type as
   above.
   Leiocarya kotschyana Hochst. in Flora **27**: 30 (1844). Syntypes from Ethiopia.
   Pollichia zeylanica (Burm. f.) F. Muell., Syst. Census Austral. Pl.: 100 (1882). Type as for *T.*
   zeylanicum.
   Boraginella zeylanica (Burm. f.) Kuntze, Rev. Gen. Pl. **2**: 435 (1891). Type as above.
   Borraginoides zeylanica (Burm.) Hiern, Cat. Afr. Pl. Welw. **1**: 720 (1898).
   Trichodesma zeylanicum forma *longifolium* Brand in Engl., Pflanzenr. **78**, [IV, 252]: 42 (1921).
   Syntypes from Tanzania and India.

Annual with a tap root, or perhaps occasionally a short-lived perennial with a
somewhat woody stock, with erect stems 0.3–1.5(2.0) cm. high, with ascending branches in
the upper part; stems clothed with short spreading hairs interspersed with coarse
tuberculate setae, or occasionally only with the tuberculate setae. Leaves (3)5–12(15) ×
1–3(4) cm., narrowly elliptic, cuneate or occasionally rounded at the base, usually acute at
the apex, sessile or subsessile, opposite or the uppermost alternate, the upper surface
covered with tuberculate setae and sometimes with smaller hairs between them, the lower
surface with tuberculate setae on the major veins and usually with a dense covering of
smaller spreading or appressed hairs between them. Inflorescence 1–5 cm. long at
flowering, elongating to up to 14 cm. in fruit, many-flowered, terminal on main and
lateral branches, bracteate for most or all their length, the bracts ovate to lanceolate and
rounded or subcordate at the base; pedicels 1–3 cm., densely clothed with conspicuous
flexuous spreading hairs up to 2 mm. long and often also with minute spreading hairs.
Sepals 7–10 × 2–3.5 mm. at flowering, enlarging to up to 20 × 6 mm. in fruit, ovate to
lanceolate, rounded at the base, acute at the apex, densely clothed with long appressed or
ascending hairs with those on the midrib and margins usually tending to be setose and
more strongly tuberculate than the others. Corolla 7–9 mm. long, slightly shorter than to
slightly longer than the sepals, the lobes broadly ovate-acuminate; lobes pale blue to lilac
or pinkish, the tube often whitish between the lobes and with red to purple markings at its
base. Fruit comprising four nutlets, falling separately to leave a pyramidal gynobase with
four strongly concave sides, each with a more or less winged margin, and a persistent
terminal style; nutlets 4–4.5 × 2.5–3.3 × 1–1.5 mm., compressed-ovoid, smooth and shiny
and usually mottled with grey and brown on the outer surface, rugose on the inner
surface.

Zambia. N: Mkupa, fl. & fr. 8.x.1949, *Bullock* 1183 (K; LISC). W: Ndola, fl. v.1961, *Wilberforce* 104
(K). C: Lusaka Distr., Mt. Makulu Res. Stat., fl. 2.vi.1956, *Angus* 1315 (K; SRGH). E: Jumbe, fl. & fr.
11.x.1958, *Robson* 55 (BM; K; LISC; SRGH). S: Mapanza Mission, fl. & fr. 14.xii.1952, *Robinson* 2 (K).
Zimbabwe. N: Hurungwe Distr., near Rukomechi R., fl. & fr. 21.viii.1959, *Goodier* 583 (K; LISC;
SRGH). W: Hwange Distr., Siatshilaba's Kraal, 24 km. NE. of Sebungwe confluence, fl. 11-16.v.1956,
Plowes 1980 (K; SRGH). E: 21 km. S. of Mutare (Umtali), fl. & fr. 1.xi.1967, *Mavi* 394 (K;
SRGH). S: Chiredzi Distr., Gona-re-Zhou, between Chitsa's Kraal and Savi-Rundi junction,
Tamboharta Pan, fl. & fr. 31.v.1971, *Grosvenor* 589 (K; SRGH). **Malawi**. N: Kasitu R. 32 km. W. of
Mzuzu, fl. 8.vii.1974, *Pawek* 8802 (K; MAL; MO; SRGH). C: Mchinji, banks of Bua R. by road to

Zambia, fl. 30.iii.1970, *Brummitt* 9549 (K; MAL). S: Mpatamanga Gorge, fl. & fr. 5.ii.1979, *Blackmore &* *Patel* 286 (K; MAL). **Mozambique**. N: Mutuali, Cotton Culture Station, fl. & fr. 16.ix.1953, *Gomes e* *Sousa* 4117 (COI; K; LISC). Z: Namagoa Plantations, fl. 3.vi.1946, *Faulkner* 297, (COI; K; PRE). T: Mazowe R., Kabankangwa Kraal, fl. & fr. 22.ix.1948, *Wild* 2595 (K; SRGH). MS: Cheringoma, Inhamitanga, fl. & fr. 29.ix.1944, *Simão* 118 (LISC). GI: Guijá. near Caniçado, fl. 21.v.1948, *Torre* 7887 (LISC). M: Goba, fl. & fr. 10.viii.1946, *Pinto* s.n. (LISC).

Widespread in tropics and subtropics of the Old World, including Australia; in eastern Africa from Sudan and Ethiopia southwards to S. Africa (Transvaal and northern Natal), but apparently absent from western Africa. In disturbed ground and waste places or sometimes in open woodland, often very abundant and a serious weed of cultivation at middle and lower altitudes; 250–1500 m.

2. **Trichodesma angustifolium** Harv., Thes. Cap. **1**: 26, t. 40 (1859). —Wright in Thiselton-Dyer, F.C. **4**, 2: 11 (1904). —Eyles in Trans. Roy. Soc. S. Afr. **5**: 457 (1916). —Friedrich-Holzhammer in Merxm., Prodr. Fl. SW. Afr. **120**: 3 (1967). —Ross, Fl. Natal: 297 (1973). —Brummitt in Kew Bull. **40**: 852, map 1 (1985).Type from S. Africa.
    Trichodesma lanceolatum Schinz in Verhandl. Bot. Ver. Brandenburg, **30**: 269 (1888). Types from Namibia.

Perennial herb to bushy subshrub with erect or decumbent-ascending stems up to 50(60) cm.; stems with short appressed or ascending tuberculate setae, the epidermis often flaking off in older plants. Leaves 2–7 × 0.1–0.6(0.9) cm., linear to linear-elliptic, narrowly cuneate at base, acute at apex, with tuberculate short appressed setae on both surfaces. Inflorescences 1–5 cm. long, with 2–8 flowers, terminal and ebracteate or the lowermost flowers sometimes solitary in leaf axils; pedicels 1–2 cm., or up to 3 cm. in fruit, clothed with appressed setae. Sepals 9–12 × 2.5–4 mm. at flowering, lanceolate, cordate at the base, long-acute at the apex, enlarging to up to 25 × 20 mm. in fruit and then ovate, strongly cordate and becoming papery, rather regularly covered by short appressed setae at all times. Corolla 12–18(22) mm. long, the tube about equalling the sepals, the ovate-acuminate lobes reflexed over the sepals, with markedly elongate apices, blue or sometimes pinkish-blue. Fruit comprising four nutlets, falling separately to leave a pyramidal gynobase, with four strongly concave sides and a persistent terminal style; nutlets 7–8 × 5–6 × c. 1mm., compressed-ovoid, the outer surface covered with conical projections c. 1.5 mm. long, each with a minute apical barb and sometimes additional reflexed barbs below the apex, the marginal projections usually coalescing at their bases to form a distinct rim.

Botswana. SW: near Tr 92, Farm 68, D'kar, fl. 16.ii.1970, *Brown* 8394 (K); 6 miles [10 km.] N. of Ramatlabama, fl. & fr. 12.xi.1977, *Hansen* 3277 (K). **Mozambique**. M: Maputo, Changalane, fl. & fr. 21.xi.1944, *Mendonça* 2991 (LISC); Sabiè, Moamba, fl. & fr. 16.x.1940, *Torre* 1800 (LISC).

Also in S. Africa (northern Natal, Transvaal and northern Cape Province) and Namibia. Stony or sandy places or open grassland; sea level to 1500 m.

Trichodesma angustifolium was recorded from Zimbabwe by Eyles in Trans. Roy. Soc. S. Afr. **5**: 457 (1916). Only one collection indicated as from there has been seen: Bulawayo, May 1898, *Rand* 388 (BM; BR), and the Eyles reference may be based on this. No collection made in this century has been traced to confirm the occurrence of the species in Zimbabwe, and some doubt is cast on the authenticity of the collection by a note on another one made later the same year by the same collector. *Rand* 636 (BM), labelled 'British Protectorate Bechuanaland, Oct. 1898' is annotated 'I have not found this species so far north as Rhodesia so far'. Bulawayo would be well outside the known distribution area of the species, and the Zimbabwe record must be considered doubtful until confirmed by a more recent collection.

3. **Trichodesma physaloides** (Fenzl) A.DC. in DC., Prodr. **10**: 173 (1846). —Gürke in Engl. & Prantl, Pflanzenfam. IV, **3a**: 99, t. 40F (1893). —Wright in Thiselton-Dyer, F.C. **4**, 2: 11 (1904). —Baker & Wright in Thiselton-Dyer, F.T.A. **4**, 2: 46 (1905), excl. syn. *T. ambacensi* Welw. —Eyles in Trans. Roy. Soc. S. Afr. **5**: 457 (1916). —Brand in Engl., Pflanzenr. **78**, [IV, 252]: 22 (1921). —Hopkins, Bacon & Gyde, Common Veld Fl.: 88, cum phot. (1940). —Weimarck in Bot. Notis. **1940**: 63 (1940). —Suessenguth & Merxm. in Trans. Rhod. Sci. Assoc. **43**: 42 (1951). —Brenan in Mem. New York Bot. Gard. **9**: 6 (1954). —Martineau, Rhod. Wild Fl.: 64, t. 22(1) (1954). —Binns, H.C.L.M.: 24 (1968). —Richards & Morony, Check List Fl. Mbala Distr.: 208 (1969). —Taton in Fl. Congon, Rwanda et Burundi, Boraginaceae: 39 (1971). —Biegel & Mavi in Wild, Rhod. Bot. Dict. ed. 2: 259 (1972). —Brummitt in Wye Coll. Malawi Proj. Rep.: 51 (1973). —Agnew, Upl. Kenya Wild Fl.: 521 cum tab. (1974). —Plowes & Drummond, Wild Fl. Rhod.: sp. 102, t. 139 (1976). —Tredgold & Biegel, Rhod, Wild. Fl.: 45, t. 29(1) (1979) pro parte, excl. t. 29(a). —Brummitt in Kew Bull. **37**: 439 (1982). TAB. **28**. Type from Sudan.
    Friedrichsthalia physaloides Fenzl in Endlicher & Fenzl, Nov. Stirp. Dec.: 54 (1839). Type as above.

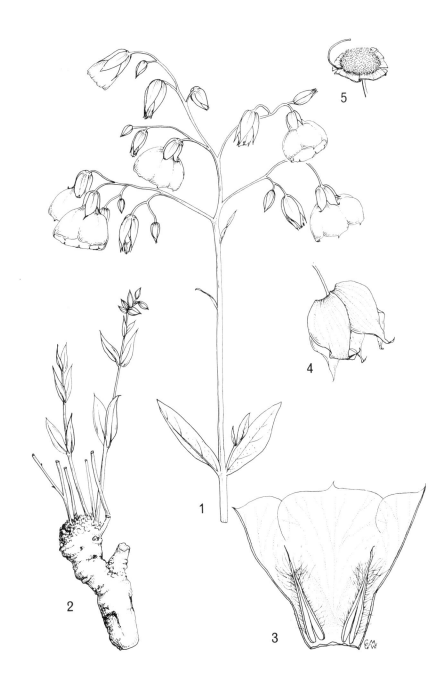

Tab. 28. TRICHODESMA PHYSALOIDES. 1, flowering stem (× ⅔); 2, rootstock and young shoots (× ⅔); 3, part of corolla with stamens (× 2); 4, fruiting calyx (× ⅔); 5, fruit with calyx removed (× ⅔), all from *Rutherford-Smith* 25.

Boraginella physaloides (Fenzl) Kuntze, Rev. Gen. Pl. **2**: 435 (1891) as *"physalodes"*. Types as above.
Borraginoides physaloides (Fenzl) Hiern, Cat. Afr. Pl. Welw. **1**: 721 (1898). Type as above.
Trichodesma droogmansianum De Wild. & T. Dur. in Bull. Soc. Bot. Belg. **39**: 69 (1900). —Baker & Wright in Thiselton-Dyer, F.T.A. **4**, 2: 47 (1905). —Brand in Engl., Pflanzenr. **78**, [IV, 252]: 23 (1921). Type from Zaire.
Trichodesma glabrescens Gürke in Engl., Bot. Jahrb. **30**: 389 (1901). Type from Tanzania.
Trichodesma ringoetii De Wild. in Fedde, Repert. Sp. Nov. Regni Veg. **13**: 110 (1914). —Brand in Engl., Pflanzenr. **78**, [IV, 252]: 22 (1921). Type from Zaire.
Trichodesma droogmansianum var. *glabrescens* (Gürke) Brand in Engl., Pflanzenr. **78**, [IV, 252]: 24 (1921). Type as for *T. glabrescens*.

Perennial herb with 1–several erect stems up to 50(60) cm. from rootstock, usually unbranched except for the inflorescence but sometimes with few sterile axillary shoots; stems ± glabrous or with sparse setae on lower internodes. Leaves (1)2–6(9) × (0.2)0.5–2.5(3.5) cm. or rarely (on sterile shoots) up to 11 × 4.2 cm., variable in shape from broadly ovate or broadly elliptic to lanceolate or linear-lanceolate or linear-elliptic, acute at the apex except in some basal leaves, cuneate to rounded at the base, ± sessile, lower ones opposite and uppermost usually alternate, the upper surface usually fairly densely clothed with tubercles which may sometimes bear minute setae, the lower surface with such tubercles inconspicuous or lacking or with them conspicuous usually only on the main veins and at the margins. Inflorescences 2–30 cm. long, with 2–9(12) primary branches each bearing 1–8(12) flowers; primary branches usually subtended by very reduced leaves but individual cymes ebracteate; pedicels 1–2.5(3.2) cm. in flower, elongating up to 4.5 cm. in fruit, glabrous or with a few tubercles just below the flower. Sepals (8)10–16(18) × (3)3.5–6(7) mm. in flower, lanceolate to sometimes ovate, enlarging to up to 27 × 22 mm. in fruit and then rounded to cordate at the base, glabrous or with tubercles at the base and round the margin and sometimes sparsely on the surfaces also. Corolla (13)15–22(25) mm. long, the lobes scarcely spreading (in herbarium specimens the sepals fully visible in flowers pressed from the side) and broadly rounded to truncate above a short apical cusp, in most of its range with a dense line of hairs down the middle of each lobe outside but sometimes glabrous, (in Zimbabwe and S. Africa more commonly glabrous); lobes white or sometimes cream or pink-tinged, the tube white with brown markings at the sinus between lobes. Fruit a single discoid nutlet 11–14 mm. diam. and 8 mm. thick, tomentose on its outer surface, with a persistent lateral style, sessile and attached by most of its lower surface to the receptacle and concealed in the accrescent calyx until its release.

Zambia. N: Mbala, fl. ix.1956, *Clayphan* 90 (K; SRGH). W: Kitwe, Ichimpi, fl. 4.ix.1964, *Mutimushi* 981 (K). **Zimbabwe**. N; Mazowe (Mazoe), fl. viii.1917, *Walters* 2312 (K; SRGH). W: Bulawayo, Circular Drive, fl. 11.x.1975, *Cross* 228 (K; SRGH). C: Harare, northern suburb, fl. 31.viii.1960, *Rutherford-Smith* 25 (K; LISC; SRGH). E: Odzani R. Valley, fl. 1914, *Teague* 267 (K; PRE; SRGH). S: Mushandike Nat. Park, 1.xi.1974, *Bezuidenhout* 102 (K; SRGH). **Malawi**. N: Nyika Plateau, valley N. of Nganda, fl.ix.1972, *Synge* WC382 (K; MAL; SRGH). C: Dedza Mt., fl. 24.ix.1969, *Salubeni* 1419 (K; SRGH). S: Bvumbwe, fl. 1.ix.1966, *Agnew* 433 (K). **Mozambique**. N: Massangulo, fl. v.1933, *Gomes e Sousa* 1478 (COI). T: between Dedza and Vila Coutinho, fl. 20.ix.1935, *Lea* 43 (K; PRE). MS: Mavita, fr. 25.x.1944, *Mendonça* 2576 (LISC).
From Sudan and Ethiopia to S. Africa (northern Natal). In grassland or woodland subject to annual burning; 900-1900 m.

4. **Trichodesma ambacense** Welw. in Ann. Cons. Ultramar, 1 [Apont. Phyto-Geogr.]: 589 (1859), as *"ambacensis"*. —Brummitt in Kew Bull. **37**: 442 (1982). Type from Angola.
Boraginella ambacensis (Welw.) Kuntze, Rev. Gen. Pl. **1**: 435 (1891). Type as above.
Trichodesma dekindtianum Gürke in Engl., Bot. Jahrb. **32**: 142 (1902); in Warburg, Kunene-Sambesi-Exped. Baum: 348 (1903). —Baker & Wright in Thiselton-Dyer, F.T.A. **4**, 2: 50 (1905). —Brand in Engl., Pflanzenr., **78**, [IV, 252]: 24 (1921). Type from Angola.
Trichodesma welwitschii Brand in Fedde, Repert. Sp. Nov. Regni Veg. **12**: 505 (1913); in Engl., Pflanzenr., **78**, [IV, 252]: 26 (1921). Type from Angola.
Trichodesma angolense Brand in Engl., Pflanzenr., **78**. [IV, 252]: 26 (1921) nom. illegit. —Taton in Fl. Congo, Rwanda et Burundi, Boraginaceae: 44 (1971). Type as for *T. ambacense*.

Subsp. **hockii** (De Wild.) Brummitt in Kew Bull **37**: 446 (1982). Type from Zaire.
Trichodesma ledermannii Vaupel in Engl., Bot. Jahrb. **48**: 529 (1912). —Brand in Engl., Pflanzenr., **78**, [IV, 252]: 24 (1921). —Heine in F.W.T.A. ed. 2, **2**: 323 (1963). Type from Cameroon.
Trichodesma hockii De Wild. in Fedde, Repert. Sp. Nov. Regni Veg. **11**: 546 (1913). —Brand in

Engl., Pflanzenr., **78**, [IV, 252]: 26 (1921). —Brenan in Mem. New York Bot. Gard. **9**: 6 (1954). —Binns, H.C.L.M.: 24 (1968). —Taton in Fl. Congo, Rwanda et Burundi, Boraginaceae: 44 (1971). —Cribb & Leedal, Mt. Fl. S. Tanzania: 57 (1980). Type as for *T. ambacense* subsp. *hockii*.

Trichodesma tinctorium Brand in Bull. Jard. Bot. Brux. **4**: 393 (1914); in Engl., Pflanzenr., **78**, [IV, 252]: 23 (1921). Type from Zaire.

Trichodesma verdickii Brand in Bull. Jard. Bot. Brux. **4**: 392 (1914); in Engl., Pflanzenr., **78**, [IV, 252]: 23 (1921). Type from Zaire.

Perennial herb with 1–several erect stems up to 50(70) cm. from rootstock, usually unbranched except for the inflorescence but sometimes with few sterile axillary shoots; stems variously setose-pubescent with irregularly spreading soft setae and often softer smaller hairs interspersed, or occasionally subglabrous with sparse stiffer tuberculate setae, very rarely subglabrous throughout. Leaves (2)4–7(9) × (0.3)0.7–3(3.6) cm. or very rarely (exceptional leaves perhaps on sterile shoots) up to 16 × 5 cm., lanceolate to broadly ovate or elliptic, acute or obtuse at the apex, cuneate to rounded at the base, sessile, the lower ones opposite, the upper sub-opposite or alternate, regularly clothed on both surfaces by either flat tubercles with or without hairs or with short setose hairs with or almost without tubercles. Inflorescences 2–30 cm. long, with 2–9(12) primary branches each bearing 1–8(12) flowers; primary branches usually subtended by very reduced leaves but individual cymes ebracteate; pedicels 1–1.2(2.7) cm. in flower, elongating to up to 3.5(4.5) cm. in fruit, with spreading hairs often hooked upwards at the apex or occasionally with only upwardly appressed hairs which may be sparse. Sepals (10)12–16(18) × (3)5–8(9) mm. in flower, ovate, usually broadly overlapping each other and margins often curved outwards, enlarging to up to 20 mm. broad in fruit and then cordate at the base, clothed with sparse to dense hairs (tuberculate or not) either closely or loosely upwardly appressed in the upper part but often spreading or reflexed towards the base, rarely subglabrous with a line of tuberculate hairs round the margin. Corolla (15)17–25(3) mm. long, the lobes spreading or reflexed (in herbarium specimens usually obscuring the sepals in a majority of fully open flowers) and triangular with an acute apex, usually glabrous but rarely with a line of hairs down the middle of the lobes; lobes usually blue (? occasionally white), the tube usually white or cream and often (? always) with brown markings at the sinus between the lobes. Fruit as in *T. physaloides*.

Botswana. N: Chobe Nat. Park, 20 km. S. of Serondella, fr. 17.x.1972, *Biegel, Pope & Russell* 4005A (SRGH). **Zambia**. B: 46 km. W. of Nangweshi, fl. 6.viii.1952, *Codd* 7416 (COI; K; PRE; SRGH). N: Isoka, fl. 24.viii.1958, *Lawton* 444 (K). W: Mwinilunga Distr., Matonchi Farm, fl. 1936, *Paterson* 8 (K). C: 29 km. NE. of Kafue R., fl. 11.vii.1930, *Hutchinson & Gillett* 3590 (BM; COI; K; LISC; SRGH). E: Katete-Chadiza, mile 26, near Nsadzu R., fl. 8.x.1958, *Robson* 22 (K; LISC; SRGH). S: Mapanza N., fl. & fr. 20.ix.1953, *Robinson* 328 (K; SRGH). **Zimbabwe**. N: Chipoli, Mowbray's Farm, fl. 3.x.1931, *Rattray* 377 (PRE; SRGH). W: Plumtree, fl. ii.1935, *McLeod* 20 (K; PRE). C: Gweru Distr., Mlezu Gov. Agric. School Farm, 28 km. SSE. of Kwekwe, fl. 23.ix.1965, *Biegel* 298 (SRGH). E: S. of Penhalonga, fr. 1934, *Gilliland* Q924 (K). **Malawi**. N: Katumbi, fr. 1.xi.1966, *Gillett* 17528 (EA; K). C: Kasungu Game Reserve, fl. 10.vii.1970, *Hall-Martin* 1736 (K; SRGH). S: Mingoli, 11 km. E. of Zomba, fr. 29.xi.1977, *Brummitt, Seyani & Dudley* 15226 (K; MAL). **Mozambique**. N: Mutuali, Nalume R. bridge, fl. & fr. 14.ix.1953, *Gomes e Sousa* 4115 (COI; K; LISC; PRE; SRGH). Z: between R. Ligonha and Malema, fr. 14.x.1949, *Barbosa & Carvalho* 4429 (K).

Subsp. *hockii* from southwards Sudan southwards to western Zimbabwe and northern Botswana, and in Nigeria and Cameroon; subsp. *ambacense* in Angola. In grassland and woodland subject to annual burning; 600–2150 m.

The above description applies only to subsp. *hockii*. Subsp. *ambacense*, confined to Angola, differs in having the leaves broadest in their upper half and long-cuneate to the base, and sepals only 3–5 mm. broad.

8. CYSTOSTEMON Balf.f.

Cystostemon Balf. f. in Proc. Roy. Soc. Edin. **12**: 82 (1883).
Vaupelia Brand in Fedde, Repert. **13**: 82 (1914).

Annual or perennial herbs or subshrubs. Leaves alternate, entire. Inflorescence a large panicle or, in species outside Flora Zambesiaca area, simple or few-flowered. Calyx 5-lobed, divided almost to the base, sometimes accrescent in fruit. Corolla tube short, cylindrical or slightly funnel-shaped, without scales in throat; lobes 5, usually much longer than the tube. Stamens 5, inserted on the corolla tube; filaments short, oblong or, in species outside Flora Zambesiaca area, swollen and narrowly elliptic to round, with a basal ciliate or hairy appendage; anthers much longer than filaments, prolonged above the thecae by laminar appendages, straight, not twisted, coherent along their margins. Ovary 4-lobed; style gynobasic, elongated; stigma capitate, small. Fruit of 1–4 nutlets attached to the flat gynobase only by the base.

A genus of about 15 species, one confined to Socotra, one in SW. Arabian peninsula, the rest in East Africa, Angola and Zambia.

1. Plants hispid; pedicels with bristles 2.0–3.5 mm. long   1. *hispidissimus* subsp. *zambiensis*
– Plants scabrous; pedicels with bristles c. 1 mm. long   -   -   -   -   -   -   - 2
2. Leaves with short patent hairs; corolla lobes 21–23 mm. long; style 26–28 mm.
     long   -   -   -   -   -   -   -   -   -   -   -   -   - 2. *mwinilungensis*
– Leaves with scattered appressed hairs (strigae); corolla lobes 14–15 mm. long; style
     17–19 mm. long   -   -   -   -   -   -   -   -   -   -   -   - 3. *loveridgei*

1. **Cystostemon hispidissimus** (S. Moore) Miller & Riedl in Notes Roy. Bot. Gard. Edin. **40**: 19, t. 6 fig. e, f (1982). Type from Angola.
    Vaupelia hispidissima S. Moore in Journ. of Bot. **58**: 49 (1920). Type as above.

An erect, perennial, several-stemmed herb up to 2.5 m. high; stems usually unbranched, more or less ribbed, scabrous to shortly hispid and with scattered spreading bristles 2–3 mm. long. Leaves sessile or subsessile, 6–12 × 1.0–1.8 cm., the superior ones progressively smaller, dense, narrowly lanceolate, hispid above and below or shaggy below, acute at apex, rounded to subcordate at base, almost leathery, with the secondary nerves not or hardly perceptible. Inflorescence a more or less dense terminal panicle with indumentum of short dense rigid hairs mixed with very numerous spreading yellowish bristles 2–3 mm. long. Bracts numerous, ovate to lanceolate, rounded to subcordate at base. Calyx densely bristly outside, sericeous inside with long ascending hairs; lobes narrowly triangular to narrowly lanceolate, 7–11 × 1.5–2.0 mm. when flower newly expanded, increasing to 15 mm. long in fruit. Corolla blue, violet or lilac; tube 6–7 mm. long, funnel-shaped, glabrous outside; lobes 12–20 × 2.5–4.0 mm., narrowly ovate-lanceolate with attenuate apices, more or less sericeous outside, mainly to the apex and with some appressed short hairs inside. Stamens inserted near the apex of the corolla tube; anthers 6–10 mm. long, minutely scabrous; terminal appendages 6–11 mm. long; filaments c. 2 mm. long with basal appendage densely hairy. Style 22–29 mm. long. Fruit usually of 1–2 well developed nutlets, the other ones abortive; nutlets c. 5 × 4 × 4 mm., ovoid, verrucose, keeled to the apex, rounded at back.

Subsp. **zambiensis** Miller & Riedl, in Notes Roy. Bot. Gard. Edin. **40**: 20 (1982). Type: Zambia, Mwinilunga Distr., between Wanulolo R. and Lunga R., *Milne-Redhead* 860 (K, holotype).*

Leaves yellowish when dry. Calyx 10–12 mm. long in flower, up to 16 mm. long in fruit. Corolla lobes 17–20 × 2.5–4.0 mm. Anthers 8–10 mm. long; terminal appendages 9–11 mm. long. Style 27–29 mm. long.

Zambia. W: Mwinilunga, fl. vi.1955, *Holmes* 1187 (K).
Only known from the Mwinilunga District. In *Brachystegia* woodland and edge of dry *Brachystegia boehmii-Monotes* scrub.
Subsp. *hispidissimus* occurs in Central and south-western parts of Angola.

2. **Cystostemon mwinilungensis** E. Martins in Garçia de Orta, Sér. Bot. **9**, 1–2: 75 (1988). TAB. **29**. Type: Zambia, 60 km. S. of Mwinilunga on Rd. to Kabompo, *Brummitt, Chisumpa & Polhill* 14110 (K, holotype).
    Cystostemon mechowii (Vaupel) Miller & Riedl in Notes Roy. Bot. Gard. Edin. **40**: 20, tab. (map) 5 & tab. 6 fig. d, h (1982) quoad specim. Zambia lecta non *Trichodesma mechowii* Vaupel (1912) nec *Vaupelia mechowii* (Vaupel) Brand (1914).

An erect perennial herb 1–2 m. high; stems somewhat weak, usually unbranched, slightly ribbed, clothed with very short patent dense rigid hairs and some scattered spreading bristles c. 1 mm. long. Leaves sessile, 8–12 × 0.9–1.4 cm. those of middle of the stem, the lower ones early deciduous, the upper ones progressively smaller, dense, narrowly oblong-elliptic to narrowly lanceolate or narrowly oblanceolate, acute at apex,

* Miller & Riedl cite inadvertently, *Milne-Redhead* 560 (K) as the holotype.

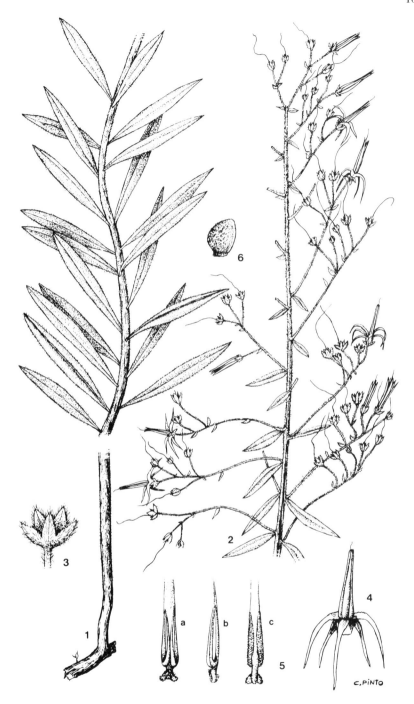

Tab. 29. CYSTOSTEMON MWINILUNGENSIS. 1, flowering stem, lower part (× ½); 2, flowering stem, upper part (× ½); 3, calyx (× 2); 4, corolla (× 1); 5, stamens in ventral (a), lateral (b), and dorsal (c) view (× 2); 6, nutlet (× 2), all from *Brummitt, Chisumpa & Polhill* 14110.

cuneate to obtuse at base, scabrous above and below but the projections smaller and erect below, larger and ascending above, sometimes bristly above; secondary nerves not perceptible. Inflorescence a large more or less lax terminal panicle with indumentum of short dense rigid hairs mixed with spreading bristles c. 1 mm. long. Bracts many, oblong-elliptic to ovate, rounded to subcordate at base. Calyx scabrous and with spreading or ascending bristles up to 1.5 mm. long outside and inside or sometimes almost sericeous inside. Lobes 4–7 × 1.5–2.5 mm. when flower newly expanded, increasing to 12.0 × 6.5 mm. in fruit, triangular-ovate to lanceolate-elliptic, acute. Corolla violet at apices of lobes and whitish at base, pale purple or bluish-mauve; tube 3.5–4.5(5.0) mm. long, widely funnel-shaped, glabrous outside, minutely hairy inside; lobes 21–23 × 3.5–4.5 mm., ovate at base, gradually attenuate into long narrowly triangular apices, more or less scabrous outside on back. Stamens inserted about 1.5 mm. from the mouth; anthers 6–7 mm. long; terminal appendages 14–15 mm. long; filaments c. 2 mm. long with basal appendage shortly and densely hairy. Style 26–28 mm. long. Fruit of 1 (? more) well developed nutlet c. 5 × 4 × 4 mm., ovoid, keeled to the apex, rounded at back, verrucose.

Zambia. W: 62.5 km. S. of Mwinilunga on Kabompo Rd., fl. 27.i.1971, *Anton-Smith* in GHS 213297 (SRGH).
Only known from Mwinilunga and Kasempa Districts in Zambia. In degraded *Cryptosepalum–Copaifera* forest and savanna woodland on Kalahari Sands.

3. **Cystostemon loveridgei** E. Martins in Garcia de Orta, Sér. Bot. **9**, 1–2: 75 (1988). Type: Zambia, Mwinilunga Distr., Muwozi stream, 67.5 km. S. of Mwinilunga on Kabompo Rd., *Loveridge* 705 (K, holotype).
Cystostemon sp. A —Miller & Riedl in Notes Roy. Bot. Gard. Edin. **40**: 21, t. 6 fig. g (1982).

An erect perennial herb or undershrub; stems somewhat weak, simple, striated, minutely roughish and with scattered strigae. Leaves sessile, 6–8 × 0.8–1.3 cm. those of middle of the stem, the inferior ones not seen, the superior ones progressively smaller and changing into bracts, narrowly lanceolate to narrowly oblanceolate, acute at apex, attenuate at base, roughish and with scattered strigae c. 0.5 mm. long above and below, almost leathery; secondary nerves not perceptible. Inflorescence a large, narrow, more or less dense terminal panicle, scabrous to bristly, the bristles ascending, to 1 mm. long or little more on pedicels. Bracts leafy, lanceolate. Calyx scabrous and with ascending bristles to 1 mm. long outside, almost sericeous inside with ascending bristles 1.5–2.0 mm. long; lobes 3.0–4.5 × 1.0–1.5 mm. when flower newly expanded, increasing to 6.5 × 2.2 mm. in fruit, narrowly triangular to narrowly ovate-lanceolate, acute. Corolla tube 3.5–4.0 mm. long, widely funnel-shaped, glabrous outside, bristly inside; lobes 14–15 × 2.5–3.0 mm., ovate at base, gradually attenuate into long narrowly triangular apices, roughish outside on back, with some short appressed bristles inside near base. Stamens inserted near the mouth of the corolla; anthers 5–6 mm. long; terminal appendages 10–11 mm. long; filaments c. 2.5 mm. long with basal appendage shortly bristly. Style 17–19 mm. long. Fruit of 1, rarely 2 well developed nutlets 3.0–3.5 mm. in diam., keeled to the apex, rounded at back, minutely verrucose.

Zambia. W: Mwinilunga Distr., Muwozi stream, 67.5 km. S. of Mwinilunga on Kabompo Rd., fl. & fr. 31.v.1963, *Loveridge* 705 (K; LISC; SRGH).
Only known from the type gathering. Ecology uncertain.

9. CYNOGLOSSUM L.

Cynoglossum L., Sp. Pl.: 134 (1753); Gen. Pl., ed. 5: 65 (1754). —Brand in Engl., Pflanzenr. **78** [IV, 252] : 114 (1921).
Paracynoglossum Popov in Komarov, Fl. URSS **19**: 717 (1953).

Annual, biennial or perennial herbs. Leaves alternate, the basal ones long petiolate, the cauline ones shortly petiolate or sessile. Inflorescence of scorpioid cymes, ebracteate or bracteate to the base. Flowers ⚥, pedicellate. Calyx 5-lobed, not or but a little accrescent. Corolla tube short, cylindrical or infundibuliform, closed with 5 scales (fornices) at the throat; lobes 5, imbricate. Stamens included; filaments very short; anthers ovate, elliptic or oblong, obtuse. Gynobase conical. Ovary of 4 separate lobes, attached to the gynobase only by a little part of the inner surface; style usually shorter than the calyx; stigma terminal, capitate or subcapitate. Fruit of 4 nutlets; nutlets glochidiate, depressed, with the outer surface flat, slightly concave or convex, the inner surface with the attach cicatrice near the apex, cohering with style or free.

A very complex group of over 60 species, distributed in the temperate and tropical regions of the world. In the Flora Zambesiaca area 8 (or more) occur, some of them probably endemic. A monographic review of the whole genus, including *Paracynoglossum* Popov. is necessary.

1. Leaves with secondary nerves subparallel and converging to the apex or only the midrib well-marked - - - - - - - - - - - - - - - - - - - 2
 – Leaves with secondary nerves merged into two submarginal well-marked nerves 4
2. Secondary nerves well-marked; cauline leaves ovate - - - - - 1. *amplifolium*
 – Secondary nerves not well-marked; cauline leaves lanceolate to oblong - - - 3
3. Pedicels at flowering c. 1.5 mm. long; corolla lobes less than 1.0 mm. long 2. *hispidum*
 – Pedicels at flowering more than 4.0 mm. long; corolla lobes more than 1.2 mm. long - - - - - - - - - - - - - - - - - 3. *aequinoctiale*
4. Fornices much broader than long, semilunar - - - - - - - - 5
 – Fornices little broader than long, ± trapez-shaped - - - - - - - 6
5. Mature nutlets up to 2.5 mm. long, immarginate, with glochids evenly dispersed over surface; cymes often divaricate; corolla light blue or white - - - - - 6. *lanceolatum*
 – Mature nutlets more than 2.5 mm. long, marginate, outer surface with glochids in the margin and in a central row and either glabrous on each side of this or sparsely glochidiate; cymes not divaricate; corolla deep blue - - - - - - - - - - 7. *geometricum*
6. Leaves narrowly lanceolate to linear-lanceolate, at middle stem up to 1.5 cm. wide - - - - - - - - - - - - - - - - 4. *coeruleum*
 – Leaves lanceolate, at middle stem usually more than 2 cm. wide - - - - 7
7. Corolla lobes white; fruiting pedicels elongated up to 8 mm. long - - - 8. *wildii*
 – Corolla lobes deep blue; fruiting pedicels not elongated - - - - 5. *inyangense*

1. **Cynoglossum amplifolium** Hochst. ex A.DC. in DC., Prodr. **10**: 149 (1846). —Gürke in Engl., Pflanzenw. Ost-Afr. **C**: 337 (1895). —Baker & Wright in F.T.A. **4**, 2: 53 (1905). —Brand in Engl., Pflanzenr. **78** [IV, 252]: 141 (1921). —Weimarck in Bot. Notis. **1940**: 65 (1940). —Robyns, Fl. Parc Nat. Alb. **2**: 132 (1947). —Taton in Fl. Congo, Rwanda et Burundi, Boraginaceae: 54 (1971). Type from Ethiopia.

A rhizomatous perennial herb 0.8–1.8(3.5) m. high, much branched, with shortly and retrorsely hispid often reddish annual stems. Inferior leaves petiolate, the superior ones sessile; petiole up to 13(42) cm. long, with sheathing base; lamina of the inferior leaves 12–20 × 7–12 cm., becoming progressively smaller to the apex, ovate or less often elliptic, shortly and sparsely hispid on both surfaces, acute at apex and sometimes apiculate, obtuse to subcordate at base and suddenly decurrent on the petiole. Inflorescence of short racemes arranged in terminal bracteate panicles; bracts ovate to lanceolate, acute. Flowers with pedicels 2–4 mm. long. Calyx appressed-pubescent outside, glabrous inside; lobes 1.0–2.5 × 1.2–1.7 mm., elliptic to very widely ovate, subacute to rounded at apex. Corolla usually blue, often with reddish throat, sometimes the whole whitish; tube 1.4–1.6 mm. long, slightly infundibuliform; limb 4–6 mm. in diam.; lobes 1.4–2.4 × 1.0–2.6 mm., very widely ovate to depressed-ovate, obtuse, much veined. Stamens inserted at 0.6–1.0 mm. from the base of the corolla tube; anthers 0.6–0.7 mm. long, elliptic-oblong, obtuse, subcordate at base; filaments 0.3–0.5 mm. long. Ovary c. 1 mm. in diam.; style overlapping the ovary by 0.3–0.6 mm.; stigma globulose, terminal. Fruits 10–12 mm. in diam., on usually reflexed pedicels up to 17 mm. long; nutlets 5–6 mm. in diam., free from the style, with the outer surface convex and ± uniformly covered with 0.5–0.6 mm. long glochids.

Zimbabwe. E: Mutare Distr., Engwa, fl. 2.ii.1955, *Exell, Mendonça & Wild* 127 (BM; LISC). **Malawi**. N: Rumphi Distr., Nyika Plateau, Lake Kaulime, fr. 16.v.1970, *Brummitt* 10797 (K). **Mozambique**. MS: Manica Distr., Serra Zuira, Tsetserra, 4 km. from the cow-house on the road to Chimoio (Vila Pery), *Torre & Correia* 15565 (COI; LD; LISC; LMU).
From Ethiopia southwards, through the high mountains of E. Africa to the Flora Zambesiaca area. On stream edges, at moist evergreen montane forest margins, in ravine forests and in montane grasslands; 1980–2440 m. (up to 3600 m. outside the Flora Zambesiaca area).

2. **Cynoglossum hispidum** Thunb., Prodr. Pl. Cap.: 34 (1794). —C.H. Wright in F.C. **4**, 2: 14 (1904). —Brand in Engl., Pflanzenr. **78** [IV, 252]: 146 (1921). —Weimarck in Bot. Notis. **1940**: 65 (1940). Type from S. Africa.

A biennial herb up to 0.6 m. high; stems erect, unbranched, hispid (retrorsely to the lower part), solitary or few arising from the remains of the rosette of old basal leaves.

Lamina 5.0–7.5(15?) × 0.8–1.3(2.5?) cm. in inferior cauline leaves, progressively smaller to the apex, not seen in basal leaves, oblong to oblanceolate, shortly hispid above and below, the bristles with bulbous base and surrounded by cystolith areas, apex acute to obtuse, base narrowed into a petiole in basal and inferior cauline leaves, obtuse in the cauline sessile ones, the margins entire; midnerve slightly depressed above, prominent below and with the bristles retrorse on basal part, the secondary nerves little perceptible or even undiscernible. Cymes arranged in a terminal thyrse. Flowers with pedicels c. 1.5 mm. long, retroflexed and elongated in fruit up to 15(25) mm. Calyx lobed to near base, lobes 1.2–2.0 × 0.6–0.8 mm., ovate-lanceolate, acute, hispid outside, glabrous inside, apparently 1-nerved. Corolla tube 1.4–1.7 mm. long, widely cylindric, glabrous; limb cleft to near base; lobes 0.6–0.8 × 0.4–0.6 mm., rounded at apex, almost veinless; fornices subquadrate. Stamens inserted at c. 0.8 mm. from the base of corolla tube; anthers c. 0.8 mm. long, ovate; filaments 0.2–0.3 mm. long. Stigma globulose, terminal on the short style. Nutlets c. 3.5 × 3 mm., ovoid, free from the style, not marginate, densely and uniformly covered with short and strong glochids.

Zimbabwe. E: Nyanga Distr., c. 22 km. from Nyanga to Rusapi, 1800 m., fl. & fr. 9.xi.1930, *Fries, Norlindh & Weimarck* 2787 (LISU).
Also in S. Africa. In grasslands along streams.

3. **Cynoglossum aequinoctiale** T.C.E. Fries in Notizbl. Bot. Gart. Berlin **8**: 416 (1923). Type from Kenya.

An erect rhizomatous perennial herb, much branched to the apex; stem and branches shortly hispid. Basal leaves many in a dense rosette, up to 19.5 × 2.5 cm., narrowly lanceolate, rough to hispid on both surfaces, acute at apex and with base narrowed into a ± long petiole. Cauline leaves fewer, mid cauline c. 4.0 × 0.8 cm., lanceolate to linear-lanceolate, acute at apex, acute or subacute at base, sessile, progressively changing into bracts to the apex; lateral nerves inconspicuous. Cymes few-flowered, arranged in an ample lax panicle. Pedicels 4.0–6.0 mm. long, swollen to the receptacle, rough-hispid, elongating up to 15 mm. and recurved in fruit. Calyx 2 mm. long, campanulate, shortly hispid outside, divided to near base; lobes c. 1.5 mm. long, ovate-lanceolate, acute to obtuse, somewhat thick and a little bent inwards. Corolla glabrous, pale-blue; tube c. 2 mm. long, cylindric-infundibuliform; lobes 1.4–2.0 × 1.4–2.2 mm., broadly ovate to widely transversely elliptic, rounded at apex, erect-spreading; fornices subquadrate, papillose, notched at apex. Stamens inserted about the middle of the corolla tube; anthers c. 0.6 mm. long, elliptic; filaments c. 0.4 mm. long. Style c. 0.6 mm. long, glabrous; stigma subcapitate. Fruit up to 10 mm. in diam.; nutlets c. 5 × 5 mm. (without glochids), ± orbicular, covered all over with glochids.

Zambia. N: Mbala Distr., 1760 m. fl. iii.1934, *Gamwell* 196 (BM).
Only 1 specimen seen from Flora Zambesiaca area. Also in Kenya. "Edge of old native garden".

4. **Cynoglossum coeruleum** Hochst. ex A.DC., Prodr. **10**: 148 (1846). —Baker & Wright in F.T.A. **4**, 2: 53 (1905). —Brand in Engl., Pflanzenr. **78** [IV, 252]: 146 (1921). —Kabuye & Agnew in Agnew, Upl. Kenya Wild Fls.: 523 (1974). Type from Ethiopia.
    Paracynoglossum afrocaeruleum Mill ex Mill & Miller in Notes Roy. Bot. Gard. Edin. **41**: 481 (1984). Type as above.
    Cynoglossum afrocaeruleum (Mill) Riedl in Linzer. Biol. Beitr. **17**, 2: 320 (1985). Type as above.

An erect or ascending perennial herb up to 1.2 m. high; stem (or stems) much branched, clothed with somewhat rigid hairs, spreading or retrorse on the lower part, appressed to the upper part as on the branches. Inferior cauline leaves long-petiolate, up to 25 × 3(5) cm., linear-lanceolate to linear-oblanceolate, sometimes narrowly lanceolate, with a petiolar portion up to 7 cm. long (radical leaves not seen). Cauline leaves at middle stem up to 14.0 × 1.5 cm., narrowly lanceolate to linear-lanceolate, hispid-pilose on both surfaces, acute at apex, acute to obtuse at base; midrib and nerves depressed above, prominent below. Cymes usually grouped by two on each peduncle, often with a flower at the fork and arranged in an ample lax panicle. Pedicels c. 2 mm. long, elongating up to 7 mm. and much recurved in fruit. Calyx c. 2 mm. long, shortly hispid outside, ciliate at the margins, devided to near the base; lobes ovate to oblong-elliptic, subacute to rounded at apex, accrescent. Corolla glabrous, blue; tube 1.5–2.0 mm. long; lobes 1.0–1.5 × 1.0–1.5

mm., circular to oblong, rounded at apex, spreading; fornices c. 0.4 × 0.6 mm., slightly bilobed, papillose. Stamens inserted about the middle of the corolla tube; anthers c. 0.5 mm. long, elliptic; filaments about as long as the anthers. Style c. 0.5 mm long; stigma subcapitate. Fruit c. 6 mm. in diam.; nutlets c. 2.5 × 2.0 mm. (without glochids), ovate, slightly marginate, covered all over with glochids.

Malawi. N: Rumphi Distr., Nyika Plateau, towards Kasaramba, c. 2.370 m., fl. & fr. immat. 28.iii.1970, *Pawek* 3387 (K).
Also recorded from Ethiopia, Kenya and Tanzania. In submontane grasslands.

5. **Cynoglossum inyangense** E. Martins in Garçia de Orta, Sér. Bot. **9**, 1–2: 76 (1988). Type: Zimbabwe, Nyanga (Inyanga) Downs, *Wild* 4934 (K, holotype).

A tall herb; stem (or stems) ascending, strong, hollow at least when dry, much branched, angular and ± laterally compressed, dark brown, retrorsely hispid below and acroscopicaly appressed hispid to the apex; branches with indumentum similar to the stem. Basal and lower leaves not seen. Mid cauline leaves sessile, up to 10.5(14) × 3.0 cm., lanceolate, acute and with an acumen 2.0–3.5 mm. long, constituted by the involute margins, obtuse to attenuate at base into a very short petiolar portion very narrowly decurrent on the stem, the upper ones much smaller, hispid both above and below but retrorsely below, at least on the proximal part, papery, brown-greenish to dark brown above, olivaceous to light brown below when dry; midrib, the 2 lateral nerves (triplinerved leaf) and the 4–7 secondary nerves on each side of the midrib impressed above and prominent below. Cymes ± lax, many-flowered, two at the ends of the branches, usually with a flower at the fork, the whole forming a large lax panicle. Pedicels 1.0–3.0(3.5) mm. long, not or little recurved in fruit. Calyx 1.5–2.5 mm. long, campanulate, scabrous to hispid or sometimes subglabrous but always ± ciliate at margins, usually divided to near base; lobes irregular in size and in shape, ovate or triangular to obovate or even spatulate, acute to rounded at apex. Corolla glabrous, deep blue; tube 1.2–1.7 mm. long, cylindrical; lobes 1.3–1.5 × 1.4–1.9 mm., subcircular to obovate-spatulate, apex rounded, with undulate margins, spreading; fornices c. 0.5 × 0.7 mm., ± trapeziform, slightly bilobed. Stamens inserted between ½ and the upper ¾ of the tube; anthers c. 0.7 mm. long, elliptic; filaments c. 0.2 mm. long. Style c. 0 7 mm. long; stigma subcapitate. Fruit c. 6 mm. in diam.; nutlets c. 3.0 × 2.5 mm. (without glochids), ovate to subcircular in dorsal view, flattened, not marginate, outer surface flat or concave, densely glochidiate and with some minute glochids mixed with the others, glochids narrowly conical the largest ones up to 0.7 mm. long, on the margin; cicatrice c. 1 × 1 mm., deltoid.

Zimbabwe. E: Nyanga (Inyanga), World's View, fl. & fr. 16.ii.1974, *Burrows* 334 (SRGH).
Appears to be confined to Nyanga region. In rocky submontane grassland; 1900–2250 m.

6. **Cynoglossum lanceolatum** Forssk., Fl. Aegypt.-Arab.: 41 (1775). —Hiern, Cat. Afr. Pl. Welw. **1**: 721 (1898). —Baker & Wright in F.T.A. **4**, 2: 54 (1905). —Weimarck in Bot. Notis. **1940**: 64 (1940). —Suesseng. & Merxm. in Trans. Rhod. Sci. Ass. **43**: 42 (1951). —Binns, H.C.L.M.: 23 (1968). —J.H. Ross, Fl. Natal: 297 (1972). —Kabuye & Agnew in Agnew, Upl. Kenya Wild Fls.: 521 (1974). TAB **30**. Type from Yemen Arab Republic.
Cynoglossum micranthum Desf., Tabl. École Bot.: 220 (1804). —Gürke in Engl., Pflanzenw. Ost-Afr. **C**: 337 (1895). —C.H. Wright in F.C. **4**, 2: 14 (1904). Type a cultivated specimen from Paris.
Cynoglossum lanceolatum subsp. *eulanceolatum* Brand in Engl., Pflanzenr. **78** [IV, 252]: 139, t. 18 fig. A–G (1921).
Cynoglossum lanceolatum subsp. *lanceolatum.* —Heine in Hutch. & Dalz. F.W.T.A. ed. 2, **2**: 324 (1963). —Taton in Fl. Congo, Rwanda et Burundi, Boraginaceae: 50 (1971).
Paracynoglossum lanceolatum (Forssk.) Mill ex Mill & Miller in Notes Roy. Bot. Gard. Edinb. **41**: 474, t. 1 fig. C, K & t. 3 fig. Ba, Bb (1984). Type as above.

A biennial or sometimes perennial herb 0.3–1.2(1.8) m. high with erect annual stems, shortly hispid with antrorse hairs toward the apex, spreading below and sometimes retrorse near the base. Inferior leaves with petiole 2.5–8(13) cm. long, the superior ones almost sessile; lamina of the inferior leaves 8–15(20) × 2–4(6.5) cm., lanceolate to oblanceolate, that of the superior ones 2–6(8) × 0.4–1.8 cm., lanceolate to sublinear, roughish to shortly hispid above, roughish to pubescent below, acute at apex, attenuate at base and decurrent on the petiole; midrib and 2 lateral veins in all leaves usually prominent below. Cymes ebracteate above the first flower, often divaricate. Flowers with

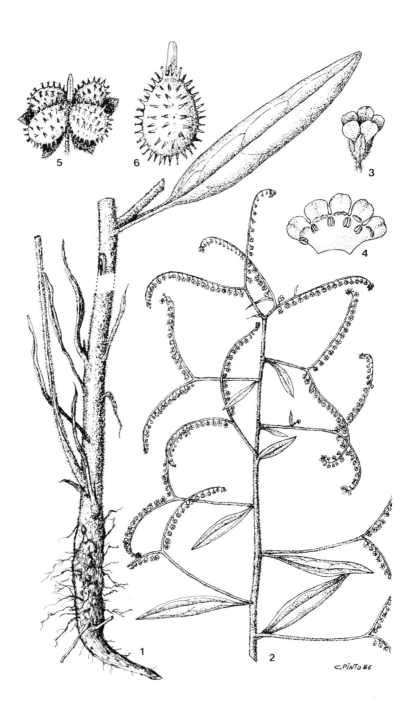

Tab. 30. CYNOGLOSSUM LANCEOLATUM. 1, stem, basal part (× ⅔); 2, stem, upper part (× ⅔); 3, flower (×5); 4, corolla opened out to show fornices and stamens (×5); 5, fruit (×5); 6, nutlet (×6), all from *Robson* 1252.

pedicels 1.0–2.5(3) mm. long, a little accrescent and recurved in fruit. Calyx 1.0–1.5 mm. long, divided to near the base; lobes ovate to oblong-ovate, roughish to shortly hispid outside, ciliate, glabrous inside, rounded at apex. Corolla 1.5–2.6 mm. long, white or blue or with the tube mauve and the limb white or bluish but deep blue at throat, glabrous; tube 0.7–1.4 mm. long, campanulate; lobes 0.6–1.5 × 0.6–1.5 mm., oblong-ovate to subcircular, rounded at apex; fornices much broader than long, semilunar. Stamens inserted at 0.6–0.9 mm. from the base of the corolla tube; anthers 0.4–0.6 mm. long, oblong; filaments 0.2–0.3 mm. long. Style thick, up to 1.5 mm. long in fruit. Nutlets 1.5–2.5 × 1.2–1.8 mm. (without glochids), ovate, united to the style by a strap which detaches itself at complete maturity ± equally covered with 0.5–1.0 mm. long glochids; outer surface flat or convex, not marginate.

Zambia. N: Kasama Distr., Kasama to Kayambi Rd., fl. & fr. 30.i.1962, *Astle* 1331 (K; SRGH). C: Mkushi, fl. & fr. 23.i.1955, *Fanshawe* 1834 (K; SRGH). **Zimbabwe**. N: Guruve Distr., Nyamunyeche Estate, fl. & fr. 12.ii.1979, *Nyariri* 680 (SRGH). W: Matobo Distr., farm Besna Kobila, fl. & fr. ii.1955, *Miller* 2709 (K). C: Marondera (Marandellas), Digglefold, fl. & fr. 7.vi.1948, *Corby* 122 (SRGH). E: Nyanga, near Cheshire, fl. & fr. 15.i.1931, *Norlindh & Weimarck* 4333 (LISU). S: Masvingo Distr., Kyle Nat. Park, Chembira Hill, fl. & fr. 22.v.1971, *Ngoni* 94 (SRGH). **Malawi**. N: Chitipa Distr., Misuku Hills, slopes above Misuku Rest House, below Mughesse Forest, fl. & fr. immat. 13.vii.1970, *Brummitt* 12088 (K). C: Dedza Mt. fl. & fr. 19.i.1959, *Robson & Jackson* 1252 (BM; K; LISC). S: Blantyre, side of Soche Mt., fl. & fr. 3.ii.1938, *Lawrence* 652 (K). **Mozambique**. N: Mandimba, serra de Massangulo, fl. & fr. 25.ii.1964, *Torre & Paiva* 10814 (C; LD; LISC; LMU; WAG). Z: Milange, fl. & fr. 11.ix.1949, *Barbosa & Carvalho* 4051 (K; LMA). T: Furancungo, monte Furancungo (Elefante), fl. & fr. immat. 15.iii.1966, *Pereira, Sarmento & Marques* 1787 (LMU). MS: Chimoio, near Garuso, fl. & fr. 25.ii.1948, *Garcia* 339 (COI; LISC). M: Namaacha, montes Ponduini, fl. & fr. 20.xi.1966, *Moura* 143 (COI; LMU).

Throughout Africa except in the northern and north-western regions, and also in Arabia. Recorded also from the mountainous parts of southern Asia. In montane grasslands, ravine forests, on river banks, stream sides, grassy hillsides, cultivated grounds and roadsides; (? 360)1000–2250 m. (up to 3000 m. outside the Flora Zambesiaca area).

7. **Cynoglossum geometricum** Baker & Wright in F.T.A. **4**, 2: 52 (1905). —Weimarck in Bot. Notis. **1940**: 65 (1940). —Binns, H.C.L.M.: 23 (1968). —Kabuye & Agnew in Agnew, Upl. Kenya Wild Fls.: 521, cum fig. (1974). —Hilliard & Burtt in Notes Roy. Bot. Gard. Edin. **43**: 348 (1986). Type: Malawi, Mt. Chiradzulu, *Whyte* s.n. (K, lectotype).
  Cynoglossum lanceolatum subsp. *geometricum* (Baker & Wright) Brand in Engl., Pflanzenr. **78** [IV, 252]: 140, t. 18 fig. H (1921). —Taton in Fl. Congo, Rwanda et Burundi, Boraginaceae: 52, t. 7 (1971). Type as above.
  Paracynoglossum geometricum (Baker & Wright) Mill ex Mill & Miller in Notes Roy. Bot. Gard. Edin. **41**: 478, t. 1, fig. d, j (1984). Type as above.

An annual or perennial herb 0.6–1.6(1.8) m. high usually much branched; stem shortly hispid, with antrorse hairs towards the apex, spreading below and very often retrorse near the base. Basal and inferior cauline leaves with petiole 3–5 cm. long, the superior ones sessile; lamina of the inferior leaves 8–18 × 2.5–7.0 cm., lanceolate, that of the superior ones 2–6 × 0.6–2.0 cm., lanceolate or ovate-lanceolate to oblanceolate, both roughish to shortly hispid on both surfaces but usually with retrorse hairs on the proximal part of the median and lower leaves, mainly on the midrib and nerves, below, apex acute, base acute to obtuse. Cymes lax, arranged in terminal few-branched panicles. Flowers with pedicels 1.5–3.0(5.0) mm. long, accrescent up to 10 mm., not or only a little recurved in fruit. Calyx 1.3–2.0 mm. long, divided to near base, lobes lanceolate to oblong-ovate, roughish to shortly hispid outside, ciliate, acute to rounded at apex. Corolla 2.5–3.5 mm. long, usually deep blue, glabrous; tube 1.2–1.8 mm. long, campanulate; lobes 1.0–2.0 × 1.0–2.0 mm., very often subcircular, sometimes transversely broadly elliptic; fornices much broader than long. Stamens inserted at 0.8–1.0 mm. from the base of the corolla tube; anthers 0.6–0.8 mm. long, oblong; filaments 0.3–0.4 mm. long. Nutlets 2.5–3.5 × 2.2–3.2 mm. (without glochids), often greyish to blackish, united to the style by a strap which detaches itself with some difficulty, well marginate, with glochids marginal and in a single central row or with few (rarely up to 4) glochids on each side of the central row.

Zambia. N: Sunzu Hill, fl. & fr. 31.iii.1960, *Fanshawe* 5604 (K). **Zimbabwe**. E: Nyanga Distr., Marora R., between Rhodes Inyanga Orchards and Hotel, fl. & fr. 22.iii.1966, *Simon* 754 (K; SRGH). **Malawi**. N: Rumphi Distr., Nyika Plateau, Kafwimba Forest, fl. & fr. 13.iv.1969, *Pawek* 2231 (K). C: Dedza Mt., fl. & fr. 20.iii.1955, *Exell, Mendonça & Wild* 1095 (BM; LISC; SRGH). S: Zomba Mt., fl. & fr. 29.vi.1955, *Jackson* 1698 (K). **Mozambique**. T: Angónia, Tsangano, Cooperativa 25 de Junho, fl. & fr. 11.vii.1979, *Macúacua & Stefanesco* 918 (LMA).

Also recorded from Ethiopia, Zaire, Uganda, Kenya, Rwanda, Burundi, Tanzania and S. Africa (Natal). In mixed deciduous woodlands, pine plantations, grasslands and on edges of evergreen forests; 1450–2300 m. (up to 3.100 m. outside the Flora Zambesiaca area).

8. **Cynoglossum wildii** E. Martins in Garçia de Orta, Sér. Bot. **9**, 1–2: 76, t. 3 (1988). Type: Zimbabwe, Mutare (Umtali), Himalayas, Engwa, *Wild* 4437 (K, holotype).

An erect biennial or perennial herb up to 0.8 m. high, unbranched or but little branched to the apex; stem subwoody below the rosette of leaves, herbaceous above, with ± scattered spreading bristles. Basal leaves (and inferior cauline ones) whithered and much eroded at flowering, with the persistent petiolar part up to 6 cm. long, hispid above and below. Cauline leaves up to 16 × 4 cm., lanceolate, acute to shortly acuminate, long and narrowly attenuate at base into a petiolar portion up to 6 cm. long, the superior ones smaller, without petiolar portion and gradually grading into bracts of the inflorescence, shortly and sparsely hispid on both surfaces and villous to the base, membranous to papery. Cymes lax, few-flowered, slender, arranged in a lax, leafy, sometimes ± corymbiform panicle. Pedicels 1.5–3.5 mm. long, elongated up to 8 mm. and somewhat recurved in fruit. Calyx 2.0–2.5 mm. long, scabrous to shortly bristly outside on the tube and on the back, ciliate at margins, divided to near the base; lobes ovate or triangular-ovate to oblong-obovate, subacute to rounded at apex, up to 3 mm. long in fruit. Corolla glabrous, white but blue at mouth; tube 1.2–1.5 mm. long, ± campanulate; lobes 1.0–2.0 × 0.8–1.5 mm., subcircular to oblong, rounded at apex, ± spreading; fornices 0.4 × 0.6 mm., ± trapeziform, papillose. Stamens inserted about the middle of the corolla tube; anthers c. 0.5 mm. long, elliptic; filaments c. 0.2 mm. long. Style c. 0.5 mm. long, up to 1.0 mm. and persisting after nutlets are shed; stigma subcapitate. Fruit c. 8 mm. in diam.; nutlets c. 3.5 × 2.5 mm. (without glochids), ovate, flattened, not marginate, not much densely covered with columnar glochids up to 0.7 mm. long, much smaller and very scattered on lower surface, with cicatrice c. 1.0 × 1.0 mm., deltoid, seemingly free from the style.

Zimbabwe. E: Mutare Distr., Himalayas, Engwa, fl. & fr. 2.iii.1954, *Wild* 4437 (K; LISC; SRGH).
Appears to be confined to the mountain range between Zimbabwe and Mozambique. Another specimen seen, *Williams* 215 (SRGH), from Mutare Distr., Vumba Mts. on streamside, seems to belong to this species. Forest floor under *Podocarpus milangianus*.

10. LITHOSPERMUM L.

Lithospermum L., Sp. Pl.: 132 (1753); Gen. Pl. ed. 5: 64 (1754).

Perennial herbs or undershrubs. Leaves alternate, entire. Flowers 5-merous, actinomorphic, isostylous or heterostylous, on leafy terminal or axillary cymes. Calyx lobed to near the base. Corolla cylindrical to infundibuliform, orange, yellowish or whitish; tube with 5 more or less glandular invaginations in the throat and with a ring of nectaries at the base, inside; lobes imbricate in bud. Stamens inserted about the middle or near the top of the corolla tube, included. Ovary with 4 free lobes, inserted on the narrow gynobase. Style gynobasic, filiform, entire or sometimes slightly 2-fid at the apex, included or sometimes almost exserted. Stigmas 2, terminal or subterminal or 1, bilobed. Fruit dry, consisting of 4 nutlets or less by abortion; nutlets detaching completely from the receptacle, ovoid to ellipsoid, erect, smooth, white and shining, with the dorsal surface convex and the ventral one obtusely angular with a flat or convex point of attachment.

A genus of about 60 species, in the temperate regions throughout the World, mainly in North America.

Lithospermum afromontanum Weimarck in Bot. Notis. **1940**: 63, fig. 7 (1940). —Hedberg in Symb. Bot. Ups. **15**, 1: 158 & 317 (1957). —Binns, H.C.L.M.: 24 (1968). TAB. **31**. Type: Zimbabwe, Nyanga Distr., foot of Inyangani Mt., *Norlindh & Weimarck* 5070 (LD, holotype).
    Lithospermum officinale sensu auct. plur. quoad pl. ex; Afr. Trop. —C.H. Wright in F.C. **4**, 2: 23 (1904). —Baker & Wright in F.T.A. **4**, 2: 59 (1905) non L.

Erect or straggling perennial herb or undershrub 0.5–1.8 m. high, much-branched above. Leaves sessile or subsessile, 35–70(95) × 5–20(27) mm., elliptic to lanceolate or linear-lanceolate, with ± appressed bulbous-based short hairs surrounded by cystoliths above, hairy below, discolorous, acute at apex, cuneate to obtuse at base, margins slightly

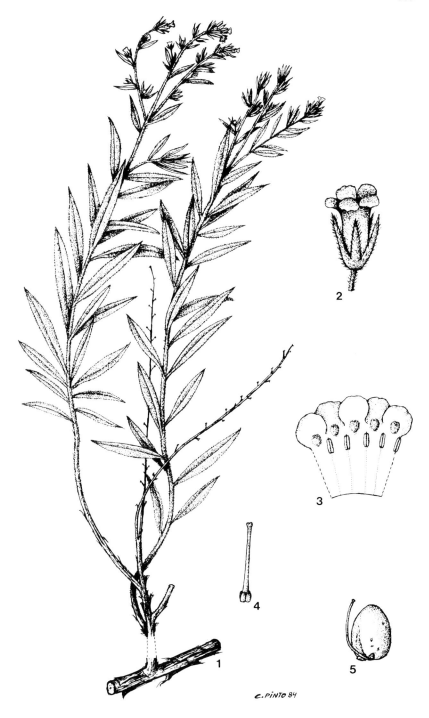

Tab. 31. LITHOSPERMUM AFROMONTANUM. 1, part of flowering branch (×⅔); 2, flower (×3); 3, corolla opened out to show stamens (×3); 4, gynoecium (×5); 5, fruit (×5), all from *Robson* 1436.

revolute, with 1–3 nerves on each side of the midrib. Cymes up to 22(30) cm. long at maturity, simple or 2-, rarely 3–4-branched, lax; bracts subsessile, similar the leaves but usually smaller. Flowers usually extra-axillary, pedicelled, heterostylous; pedicels 2–3 mm. long, up to 8 mm. long on the fruit. Calyx 2.5–5.0 mm. long, up to 10 mm. long on the fruit, bristly hairy outside, strigose inside; lobes linear to lanceolate, acute, subequal or unequal. Corolla infundibuliform, appressed hairy outside, yellowish; tube 4.5–7.0 mm. long, glabrous or glandular hairy inside; glandular invaginations in the throat ± trapeziform; nectaries at the base of the tube 10, free or ± united; corolla lobes 1.5–3.5 × 1.5–3.5 mm., circular to oblong, rounded at apex, papillose inside, margin more or less undulate, spreading. Stamens inserted above the middle of the corolla tube, often near the throat; anthers 1.0–1.5 mm. long, oblong, rounded at the ends, sometimes apiculate; filaments 0.5–0.8 mm. long, glabrous. Ovary c. 0.6 mm. wide, glabrous; style 1.2–1.6 mm. long in short-styled flowers, c. 3 mm. long in medium-styled ones and c. 7 mm. long in the long-styled ones; stigmas 2, terminal or subterminal. Fruit often with 1 nutlet, less often with 2 or 3; nutlets c. 3.5 × 2.5 mm., ovoid, rounded dorsally and obtusely keeled ventrally.

Zambia. E: Nyika, fl. & fr. 29.xii.1962, *Fanshawe* 7290 (K; SRGH). **Zimbabwe**. E: Mutare Distr., Engwa, fl. & fr. 10.ii.1955, *Exell, Mendonça & Wild* 351 (BM; LISC; SRGH). **Malawi**. N: Nyika Plateau, Chelindi Hill, fl. & fr. 2.i.1976, *Phillips* 868 (SRGH). C: Chongoni Mt., fl. & fr. 3.ii.1959, *Robson* 1436 (K; LISC; SRGH).

Also occurring in Sudan, Ethiopia, Zaire, Uganda, Kenya, Tanzania and S. Africa (Natal). In grasslands, low scrub and montane forests, often on rocky soils and in ravines; 1370–2.300 (3.950) m.

The small number of specimens seen make it impossible to establish with certainty the type of heterostyly in this species. Indeed, the length of style attributed to the short-styled flowers was seen only in the material of the type collection *Norlindh & Weimarck* 5070 (K; LD, holotypus n.v.), while the length of style attributed to the long-styled ones was seen only in the material of the collection *Exell, Mendonça & Wild* 351 (BM; LISC; SRGH). I do not perceive any correlation between the length of style and that of the stamens or with their position.

INDEX TO BOTANICAL NAMES

NOTES

NOTES

NOTES

NOTES

NOTES

NOTES

NOTES

NOTES